DIGITAL AND DATA COMMUNICATIONS

DIGITAL AND DATA COMMUNICATIONS

VINCENT F. ALISOUSKAS
DeVry Institute of Technology

WAYNE TOMASI
DeVry Institute of Technology

PRENTICE-HALL, INC., Englewood Cliffs, N.J. 07632

Library of Congress Cataloging in Publication Data

Tomasi, Wayne.
  Digital and data communications.

  Includes index.
  1. Data transmission systems.   2. Digital
communications.   I. Alisouskas, Vincent F.
II. Title.
TK5105.T66 1985   621.38'0413   84–18115
ISBN 0–13–212424–6

Editorial/production supervision and
    interior design: *Ellen Denning*
Cover design: *Edsal Enterprises*
Manufacturing buyer: *Gordon Osbourne*

Printed in the United States of America

10 9 8 7 6

ISBN 0-13-212424-6 01

PRENTICE-HALL INTERNATIONAL, INC., *London*
PRENTICE-HALL OF AUSTRALIA PTY. LIMITED, *Sydney*
EDITORA PRENTICE-HALL DO BRASIL, LTDA., *Rio de Janeiro*
PRENTICE-HALL CANADA INC., *Toronto*
PRENTICE-HALL HISPANOAMERICANA, S.A., *Mexico*
PRENTICE-HALL OF INDIA PRIVATE LIMITED, *New Delhi*
PRENTICE-HALL OF JAPAN, INC., *Tokyo*
PRENTICE-HALL OF SOUTHEAST ASIA PTE. LTD., *Singapore*
WHITEHALL BOOKS LIMITED, *Wellington, New Zealand*

CONTENTS

PREFACE

Since 1973, technological advances in the digital area have transformed telecommunications into a highly dynamic field. Where previously telecommunication systems had only voice to accommodate, with the advent of the microprocessor chip and the accompanying low-cost computers, the need for transmission of digital information has arisen. Digital technology has also provided a means of building smaller, cheaper, and more reliable telecommunications equipment. Laser technology is providing a means of transmitting information on optic fibers. The introduction of satellites into space has expanded long-distance communications.

Along with these advances, numerous texts have been published about each new area. Because of the complexity of each area, many of these texts are highly specialized. Other texts presume either that the reader has already been in the communications field and wishes to update his or her knowledge, or that the reader is highly skilled in mathematics and can follow complex derivations that establish limits on equipment and techniques or derive the probabilities associated with error rates. These books are excellent texts for the audience intended.

However, the rapid advances in technology in this area has aroused the interest of many who have never been previously associated with this field. The unique terms and numerous acronyms used in communications, if not explained, can leave the newcomer bewildered. Also, just as the advances in digital technology have brought about the introduction to digital theory and computer programming at the high school level, the authors foresee the introduction of telecommunications systems at lower levels in our educational institutions. This is the goal of this book—to provide an understanding of existing systems to a newcomer in this area. Hopefully, this book is written so that a reader with a background in basic communications (AM and FM theory), basic digital theory, and mathematics through trigonometry can easily become acquainted with this area. Numerous examples are provided for a better understanding of the theory.

Since the existing telephone lines are still the major media of all communication systems, Chapter 1 explains how direct calls are made (Direct Distance Dialing) and describes some of the associated hardware and terminology. It also describes the private-line services that are available to the subscriber. Chapter 2 is a general chapter which provides an overview of the various aspects involved in a communications system: whether it is two-point or multipoint; if multipoint, what configuration; whether it is synchronous or asynchronous; line protocol; and error detection. Many of these aspects are covered in detail in subsequent chapters. Chapter 3 deals with the telephone lines themselves and describes the specifications they must meet in order to conduct information. It also describes the various forms of distortion that may be produced in a transmitted message. If messages are to be transmitted in digital form, Chapter 4 describes the various codes that are used and error detection and correction schemes. Chapter 5 explains the operation of a UART, a device whose main purpose is to convert parallel digital information into serial format, and vice versa. Information coming from a data terminal is in parallel form. It must be converted to a serial format in order to be sent on a single transmission line. If the data terminal is receiving information, the reverse must be accomplished. Chapter 6 deals with the RS 232C interface signals between the UART and the modem. RS 449, RS 422, and RS 423, the anticipated replacements for the RS 232C, are also described. Chapter 7 deals with the various modulation schemes used to transform digital signals to analog form suitable for transmission on a voice line. These digital signals which have been converted to analog form can now be frequency-division multiplexed with other voice channels and the resultant signal transmitted on an analog line. This is the subject of Chapter 8. Since there are lines suitable for the transmission of digital information, Chapter 9 explains pulse code modulation and delta modulation—schemes for converting analog signals to digital pulses. It also explains how these channels are time-division multiplexed together prior to being transmitted on a digital carrier. Chapters 10 through 12 deal with line protocol. Chapter 10 describes asynchronous protocol, while Chapters 11 and 12 describe the synchronous protocols: Bisynch, SDLC, and HDLC. Bisynch is a character-oriented protocol, while SDLC and HDLC are bit-oriented protocols. Chapter 13 describes how different types of computers and terminals are connected to form local area networks, and packet switching and the associated standards governing their use.

VINCENT F. ALISOUSKAS
WAYNE TOMASI

1

THE TELEPHONE NETWORK

The American Telephone and Telegraph Company (AT&T) describes the Bell System as "the world's most complicated machine." A telephone call made from any telephone in the United States to any other telephone must use this machine. The characteristic of the Bell System that makes it unique from other giant corporations is that every piece of equipment, technique, or procedure, new or old, must be capable of working with the rest of the system. This is equivalent to interfacing every IBM computer with every other IBM computer or manufacturing parts for 1983 Buicks that would also work in 1952 Chevrolets. The Bell System, prior to January 1, 1983, serviced 83% of the telephones in the United States. It was the world's largest corporation, employing over 1,000,000 people. Its assets were three times those of General Motors; its revenues 20 times those of Bank of America. What was previously the Bell System, in conjunction with nearly 2000 independent telephone companies, make up the Public Telephone Network (PTN). The PTN was designed for voice communications. When the high-volume data communications boom hit in the early 1970s, the use of the PTN was an attractive alternative to constructing separate facilities (at tremendous cost) for data circuits. Gradually, communications systems are being developed that use fiber optic links and satellite transponders which are designed for data transmission. It will be many years before these systems provide any real relief to the already overloaded PTN.

The following explanation is limited to telecommunications using the PTN. *Telco* includes all of the telephone companies that make up the PTN. Telco offers two general categories of service: *direct distance dialing* (DDD) and *private line.* DDD originally included only those switches and facilities required to complete a long-distance telephone call without the assistance of a Telco operator. The DDD now includes the entire public switched network; that is, any service associated with a telephone number. Private-line services are dedicated to a single user.

DDD NETWORK

The DDD network, commonly referred to as the *dial-up* or *switched network*, imposes several limitations that must be overcome before it can be used for data communications. A basic understanding of the electrical operation of the telephone network would be helpful. The DDD network can be divided into four main sections: instruments, dial switches, local loops, and trunk circuits. An *instrument* is the device used to originate and receive signals, such as a telephone set (telset). The instrument is often referred to as *station equipment* and the location of the instrument as the *station*. A *dial switch* is a programmed matrix that provides a temporary signal path. A *local loop* is the dedicated transmission path between an instrument and the nearest dial switch. A *trunk circuit* is a transmission path between two dial switches. The dial switches are located in Telco central offices and are categorized as local, tandem, or toll. A *local* dial switch serves a limited area. The size of the area is determined by how many telephone numbers are required or desired in a given geographical area. Telco designates these areas as branch area exchanges. A branch exchange is a dial switch. The *subscriber* is the operator or user of the instrument: if you have a home telephone, you are a subscriber. A subscriber is the customer of Telco: the person placing the call. Telco refers to this person as either the talker or the listener, depending on their role at a particular time during a conversation.

A telephone number consists of seven digits. The first three digits make up the prefix while the last four constitute the extension number. Each prefix can accommodate 10,000 telephone numbers (0000 to 9999). The capacity of a dial switch is determined by how many prefixes it serves. The local dial switch provides a two-wire cable (local loop) for each telephone number it serves (Figure 1–1). One wire is designated the *tip* and the other the *ring*. The station end of the loop is terminated in a telephone set. The dial switch applies -48 V dc on the tip and a ground on the ring of each loop. This dc voltage is used for supervisory signaling and to provide a talk battery for the telset microphone. On-hook, off-hook, and dial pulsing are examples of supervisory signaling.

When a subscriber goes off-hook (lifts the handset off the teleset cradle), a switch hook is released, completing a dc short between the tip and the ring of the loop through the telset microphone. The dial switch senses a dc current in the loop and recognizes this as an off-hook condition. This procedure is referred to as a *loop start operation*: the loop is completed to indicate an off-hook condition. The dial switch responds with an audible dial tone. On hearing the dial tone, the subscriber dials the destination telephone number. The originating and destination telephone numbers are referred to as the *calling* and the *called numbers*, respectively.

Dialing is accomplished with switch closures (dial pulses) or touch-tone signaling. Dial pulsing is the interruption of the dc loop current by a telset dialing

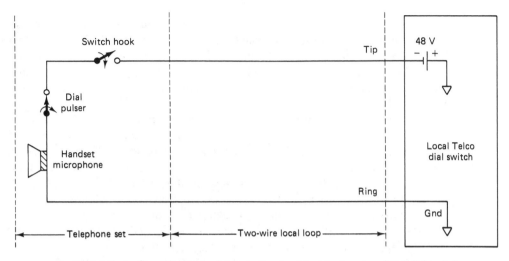

Figure 1-1 Simplified two-wire loop showing telset hookup to a local dial switch (loop start operation).

mechanism. Eight tone oscillators are contained in each telset equipped with a touch-tone pad. In touch-tone signaling, depending on the digit depressed, the telset outputs two of the eight tone frequencies.

After the entire called number has been dialed, the dial switch searches for a signal path through the switching matrix to the loop associated with the number called. Once the dial switch identifies a signal path and locates the destination loop, it tests the loop for an off-hook (busy) condition. If the destination loop is busy, the dial switch signals the calling loop with a busy signal. A station busy signal is a 60-ppm buzz. If the destination loop is on-hook (idle), the dial switch applies a 20-Hz 110-V ac ringing signal to it. A typical ringing cycle is 2 seconds on, 4 seconds off. When the destination telephone is answered (goes off-hook), the dial switch terminates the ringing signal and completes the transmission path between the two loops through the matrix. The signal path through the dial switch will be maintained as long as both loops remain closed. When either instrument goes on-hook, the signal path is interrupted.

What if the calling and the called telephone numbers are not served by the same dial switch? Generally, a community is served by only one local telephone company. The community is divided into zones; each zone is served by a different dial switch. The number of zones established in a given community is determined by the number of stations served and their density. If a subscriber in one zone wishes to call a station in another zone, a minimum of two dial switches are required. The calling station receives off-hook supervision and dial pulses are outputted as previously described. The dial switch in the calling zone recognizes that the prefix of the destination telephone number is served by a different dial switch. There are two ways that the serving dial switch can complete the call. It can locate a direct trunk (interoffice) circuit to the dial switch in the destination zone or it can route the call through a tandem switch. A *tandem switch* is a switcher's switch. It is a switching matrix used to interconnect dial switches. Trunk circuits that terminate in tandem switches are called *tandem trunks*. Normally, direct trunk circuits are provided only between adjacent zones. If a call must pass through more than one zone, a tandem switch must be used (Figure 1–2). If no direct trunks between the originating and the terminating dial switches exist and a common tandem switch is not available, the call is classified as a *toll call* and cannot be completed as dialed. Toll calls involve an additional charge and the dialed number must be preceded by a "1."

The telephone number prefix identifies which particular dial switch serves a station. With three digits, 1000 (000 to 999) prefixes can be generated for dial switches. A single metropolitan exchange alone may serve 20 to 30 prefixes. In the United States, there are over 20,000 local dial switches. It is obvious that further encoding is required to differentiate between the same prefix and extension in two different parts of the country. In the United States, an additional three-digit area code is assigned to each telephone number. The area code precedes the prefix but needs to be included only when calls are destined outside the area of the originating station. When a dialed telephone number is preceded by a "1," it is routed from the local dial switch to a toll switch by way of a *toll-connecting trunk*. Telco's present toll-switching plan

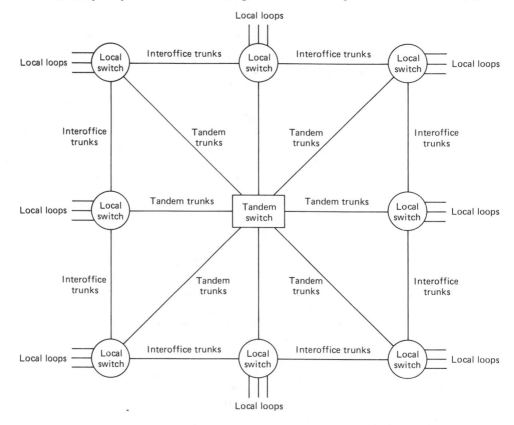

Figure 1–2   Community telephone system showing the use of a tandem switch to facilitate interzone calling.

includes five ranks or classes of switching centers. From highest to lowest classification, they are the regional center, sectional center, primary center, toll center, and end office. Local dial switches are classified as end (central) offices. All toll switches are capable of functioning as tandem switches to other toll switches.

The Telco switching plan includes a switching hierarchy that allows a certain degree of route selection when establishing a long-distance call (Figure 1–3). The choice is not offered the subscriber, but rather, the toll switches, using software translation, select the best route available at the time the call is made. The best route is not necessarily the shortest route. It is the route requiring the fewest number of dial switches. If a call cannot be completed because the necessary trunk circuits are not available, the local dial switch signals the calling station with an equipment busy signal. An equipment busy signal is similar to a station busy signal except that it repeats at a 120-ppm rate. The worst-case condition encountered when completing a long-distance call is when seven tandem toll (intertoll) trunks are required. Based on Telco statistics, the probability of this occurring is 1 in 100,000. Because software translations in the automatic switching machines permit the use of alternate routes, and each route includes many different trunk circuits, the probability of using the same facilities on identical calls is unlikely. This is an obvious disadvantage when using the PTN for data transmission because inconsistencies in the transmission parameters are introduced

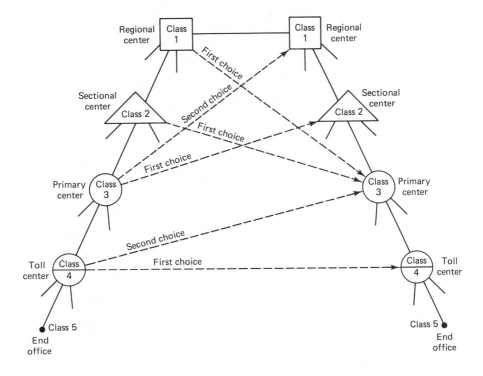

Figure 1–3 Public telephone switching hierarchy showing some possible route choices to complete a toll call between two end offices.

from call to call. Telco guarantees the transmission parameters of each local loop and trunk circuit to exceed the minimum requirements of a basic voice-grade (VG) communications channel. However, two to nine separate facilities in tandem may be required to complete a telephone call (see Figure 1–4). Since transmission impairments are additive, it is quite possible that the overall transmission parameters of a telephone connection, established through the public switched network, may be substandard. Because transmission paths vary from call to call, it is difficult to compensate for line impairments. Subscribers to the DDD network lease a dedicated loop from their station to the nearest Telco dial switch. Any additional facilities required for the subscribers to complete a call are theirs only temporarily. The subscriber uses these facilities only for the duration of the call and then they are made available for other users of the network. These temporary facilities are called *common usage trunks*—they are shared by all of the subscribers of the network.

The switching transients associated with the dial switches are another disadvantage of using the public switched network for data transmission. The older dial switches were electromechanical machines. Mechanical relay contacts were used to establish a signal path. The contact closures in the switching machines induced static interference that bled into adjacent signal paths. The static electricity caused impulse noise which produced transmission errors in the data signals. Telco is rapidly converting to *Electronic Switching Systems* (ESS). ESS machines are by no means perfectly quiet, but they are a tremendous improvement over the older electromechanical machines.

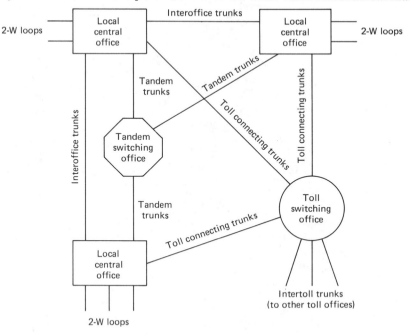

Figure 1–4 Typical switching layout showing relationship between local, tandem, and toll switches in the public telephone network.

In order for a toll call to be completed, the dialed phone number must be transferred from switch to switch. Ultimately, the switching matrix at the destination end office requires the prefix and extension numbers to establish the final connection. The transmission paths between switching machines are very often carrier systems: microwave links, coaxial cables, or digital T-carriers. Microwave links and coaxial cables use analog carriers and are ac coupled. Therefore, the traditional dc supervisory and dial pulsing techniques cannot be used. An alternate method of transferring supervisory signals using a *single-frequency (SF) tone* has been devised. An idle trunk circuit has a 2600-Hz SF tone present in both directions. An off-hook indication at either end is indicated by the removal of the SF tone. The receiving switch acknowledges the off-hook indication by removing the SF in the opposite direction. Two methods are available for dial pulsing. The SF tone can be pulsed on and off to represent the dialed number or a signaling method called *multifrequency (MF) signaling* can be used. MF is a two-of-six code similar to touch tone. However, the MF tone frequencies are higher and are transmitted at a faster rate. Touch tone and MF are not compatible. Digital T-carriers use a completely different method for transferring supervisory information. This method is explained in Chapter 9.

Hybrids and echo suppressors.   Local loops (dial-up lines) consist of a two-wire pair (signal and ground). However, to achieve satisfactory long-distance performance, the telephone company found it necessary to provide four-wire circuits (see Figure 1–5). When two-wire loops are switched onto four-wire facilities as in a long-distance telephone call, an impedance-matching device called a *hybrid* or *terminating set* (Figure 1–6) is used to affect the interface.

The hybrid coil transfers the signal from the two- to the four-wire line. The balancing network compensates for impedance variations in the two-wire portion of the circuit. The amplifiers and attenuators adjust the signal voltages to required levels. Signals traveling W-E enter the terminating set from the two-wire loop and are inductively coupled into the W-E portion of the four-wire trunk. Signals received from the E-W portion of the four-wire trunk are applied to the center taps of the hybrid coils. If the impedance of the two-wire loop and the balancing network are properly matched, all currents produced by the E-W signal in the upper half of the hybrid coil will be equal and in opposite directions. Therefore, any voltages induced in the secondaries will cancel. This prevents any received signal from being returned to the sender.

If the impedance of the two-wire loop and the balancing network are not matched, the voltages induced in the secondaries of the hybrid coil do not cancel completely. This imbalance causes a portion of the received signal to be returned to the sender on the W-

Figure 1–5   Simplified diagram of a long-distance telephone connection.

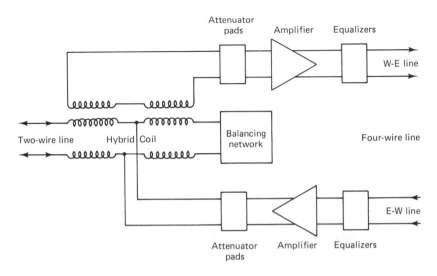

Figure 1–6   Two wire-to-four wire terminating set (hybrid).

E portion of the four-wire trunk. The returned portion of the signal is heard as an echo by the talker and, if the round-trip delay of this signal exceeds 45 ms. the echo can become quite annoying. Delays of this magnitude occur when the distance between the two telephones exceeds 1500 miles. *Echo suppressors* (Figure 1–7) are inserted at one end of a four-wire trunk circuit to eliminate this echo. The speech detector in the echo suppressor senses the presence and the direction of the signal. It then enables the appropriate amplifier and disables the amplifier for the opposite direction. With an echo suppressor in the circuit, the two parties cannot talk simultaneously. If the conversation is changing rapidly, the people talking may hear the echo suppressor turning on and off.

   The telephone lines are not only used for voice but also for the transmission of digital information. When used in this manner (Chapter 9), simultaneous transmission of digital information in both directions is often desired. To accommodate this form of transmission, the telephone company has equipped the echo suppressors with a disabling mechanism. A single-frequency tone in the range 2010 to 2240 Hz, if applied for a period of 400 ms or longer, will disable the echo suppressor—both amplifiers will be enabled. The echo suppressor will remain disabled as long as signals in the 300- to 3000-Hz band are transmitted in either direction. If this signal is interrupted for 50 ms or more, the echo suppressor will automatically become active again.

PRIVATE-LINE SERVICE

In addition to subscriptions to the public switched telephone network, Telco offers a comprehensive assortment of private-line services. Private-line subscribers lease those facilities required for a complete circuit. These facilities are hard-wired together in the Telco offices and are available only to one subscriber. Private-line circuits are

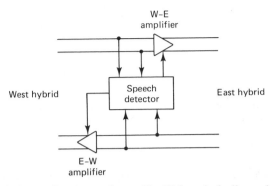

Figure 1-7 Echo suppressor.

dedicated, private, leased facilities. A dedicated circuit can be designed to meet voice-grade requirements from station to station—the end-to-end transmission parameters are fixed at the time the circuit is installed and will remain relatively constant. Circuit impairments will also remain relatively constant and can be compensated for by the subscriber. Private-line circuits afford several advantages over conventional dial-up circuits:

1. Availability
2. Improved performance
3. Greater reliability
4. Lower cost

Since private-line circuits are leased on a 24-hour basis, they are always available to the subscriber. Because the transmission parameters on a private-line circuit are guaranteed end to end, the overall performance is improved and a more reliable communications link is established. Heavy-usage private-line circuits are more economical than dial-ups. However, dial-ups are more cost-effective for a person using the lines only a small percentage of the time.

Private-line circuit arrangements differ from dial-ups only in the fact that their circuits are permanently connected and dial switches and common-usage trunks are unnecessary. The terms "local loop" and "trunk" have a slightly different meaning to private-line subscribers. On a private-line circuit, a local loop is a transmission path between an instrument and the nearest Telco office; a trunk circuit is a transmission path between two Telco offices (see Figure 1-8). The only change in the definition is the substitution of "Telco office" for "dial switch."

Examples of private-line offerings:

1. Foreign exchange (FX)
2. Full data (FD)
3. Full period (FP)
4. Digital data service (DDS)

FX circuits differ from conventional DDD subscriptions only in the fact that subscribers, instead of leasing a dedicated loop to the nearest dial switch, lease a loop to a dial switch of their choice. This facilitates toll-free interzone calling to specific zones.

Figure 1-8 Simplified private-line circuit layout showing dedicated loops and trunks with hard-wired cross-connects in each Telco office (two-point circuit).

FD circuits are four-wire, dedicated data circuits capable of full-duplex operation at a data rate of 9600 bps (bits per second). A local bank's system of automatic teller machines is an example of an FD circuit. FP circuits are four-wire, dedicated voice circuits. The hoot-and-holler (yell-down) circuits used by auto dismantlers (previously called "junkyards") to locate used auto parts is an example of an FP circuit. DDS circuits provide two-point, full-duplex operation at synchronous data rates of 2.4, 4.8, 9.6, or 56 kbps. DDS is intended to provide a communications medium for the transfer of digital data from station to station. Conventional digital-to-analog data modems are replaced by digital-to-digital communications service units (CSUs). DDS circuits are guaranteed to average 99.5% error-free seconds at 56 kbps, and even better performance is achieved at the lower bit rates.

QUESTIONS

1. Identify the four main sections that make up the DDD network.
2. What determines the capacity of a dial switch?
3. List three examples of supervisory signals.
4. Explain loop start operation.
5. Identify two methods by which dialing may be accomplished with a telephone set.
6. What is a tandem switch called?
7. How many unique telephone numbers can a three-digit prefix accommodate?
8. Identify when the three-digit area code must be included when dialing a number.
9. How does a local dial switch identify a toll call?
10. List the five classes of Telco's present switching plan.
11. The best route between an originating and a destination station is always the shortest route. (T, F)
12. Explain the difference between an equipment busy signal and a station busy signal.
13. What is the predominant disadvantage of using the DDD network for data transmission?
14. Identify two methods of transferring supervisory signals over analog carrier systems.
15. Local loops are two-wire facilities. (T, F)
16. Identify the Telco device that performs two- wire to four-wire conversion.
17. What is the purpose of an echo suppressor?
18. When are echo suppressors required?
19. Private-line circuits are dedicated to a single user. (T, F)
20. Identify four advantages of a private-line circuit compared to a dial-up circuit.

2

COMMUNICATIONS SYSTEM OVERVIEW

Communications is the transmission of information from one place to another. The transmission path may be short, as when two people are talking face to face with each other or when a computer is outputting information to a printer located in the same room. *Telecommunications* is long-distance communications. The original source information is in either analog (voice) or digital form. Voice frequencies are either transmitted directly on a voice band communications channel or used to modulate a carrier frequency and then the modulated waveform is transmitted. Conversational voice frequencies are normally in the range 300 to 3000 Hz. The voice channel over which these frequencies are transmitted has an allocated bandwidth of 4 kHz (0 to 4 kHz). Voice source information can also be transmitted in digital format. *Codecs* (coders/decoders) are integrated circuits which are capable of converting voice frequencies to a series of digital pulses or taking the digital pulses and converting them back to voice frequencies. Codecs or similar circuits perform the necessary conversions at the transmit and receive ends and the digital pulses themselves are transmitted on special lines called digital T-carriers. Digital source information can take on one of two forms. It can be raw digital data such as the op codes, addresses, and data of a computer program, or it can be a computer-contained message which is stored and/or transmitted in a digital code such as ASCII or EBCDIC. Digital information is characterized by having all information represented by a sequence of two voltage levels which represent logic 1's and 0's. In general, *digital communications* is the transmission of information in digital form (1's and 0's). The source information may be either digital or analog. *Data communications* is the transmission of information that was originally digital in nature. This information can be transmitted as either digital or analog signals. Figure 2–1 illustrates the distinction between these two forms of communications.

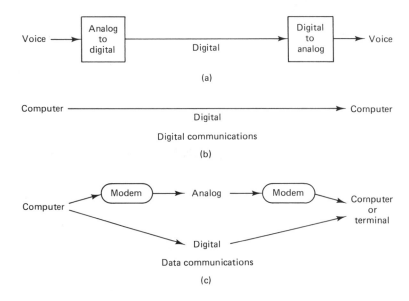

Figure 2-1   Digital and data communications.

These definitions are very broad and, as will be seen in Chapters 7 and 9, the distinction is not that exclusive.

Both the computer system and the data communications system can be very simple or very complex. The data communications system may simply be a link between a computer and a remote terminal, or it may be a link between a central computer and many terminals. Banks are good examples of such a system where all branch locations in a city tie into one main computer. These communications systems are broadly categorized as *two-point* or *multipoint* systems. A two-point system can be between a computer and a printer and/or a CRT. In such a two-point system, the flow of information would always be unidirectional. A two-point system could also be between two computers that have keyboards, printers, and displays as peripheral equipment. In this system, bidirectional transfer of information is possible. The computer which initiates information transfer is called the *master*, while the other computer is called the *slave*. Obviously, these roles are interchangeable depending on which computer initiates the call. In such a system, a problem of *contention* arises when both computers attempt to initiate a call simultaneously. Should this be possible, built-in delays resolve the problem and give one priority over the other.

Multipoint links take on many variations; the network selected is determined by system requirements. A few of these variations are depicted in Figure 2-2. The *star network* has the advantage of ready access by remote sites to the central computer. The associated disadvantage is the tariff paid for the separate lines for each terminal. If the line traffic between the remote terminals and the central computer is small, there may not be sufficient justification for such a system. But then, the ready access may be an overriding factor. In a *ring network*, the information is passed around in a loop. Each terminal extracts and inserts its own information. In this type of configuration, if one

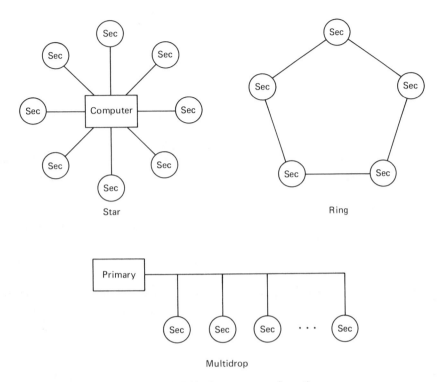

Figure 2–2 Multipoint system configurations.

element in the loop goes out of commission, the entire system is down. Backup configurations are possible where information can be circulated either clockwise or counterclockwise in the loop. Probably the most used configuration is the *multidrop*. Here, all of the terminals are connected to the same main line. In this system, the main computer is identified as the *primary* (host, CPU) and the connecting terminals are called *secondaries* (remotes, tributaries). The primary is so called because it controls all movement of information. A secondary cannot transmit or receive information unless it is allowed to do so by the primary. The primary can communicate with all of the secondaries, but each secondary can communicate only with the primary. If one secondary wishes to send a message to another secondary, it must send the information to the primary, which would then relay the information to its intended destination. In an expanded system, one station can be a secondary in one data link and serve as a primary in a different data link. In Figure 2–3, station D is a secondary in a link for which A is the primary. Station D is also the primary for the secondary stations E, F, and G. In this text, a multipoint system will be used synonymously with a multidrop system.

The types of communications possible between any two points are classified as:

1. *Simplex:* Information can be sent only in one direction. If errors are present in a transmission, the receive end has no way of asking for a retransmission. UPI and AP have such a system where the central news-gathering facility transmits the

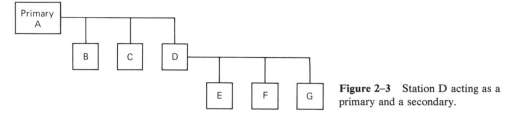

Figure 2–3 Station D acting as a primary and a secondary.

news to many regional offices. These offices are equipped with read-only printers (ROPs). The printers can display received information but cannot transmit. Simplex systems require two-wire lines (signal and reference).

2. *Half-duplex* (HDX):   Information can be transmitted in either direction but not simultaneously. Communicating with a walkie-talkie is an example of such a system. These transceivers have a push-to-send or a send/receive button. When in the send mode, the receiver is disabled; when in the receive mode, the transmitter is disabled. HDX systems also require only two-wire lines.

3. *Full duplex* (FDX) or *duplex:*   Information can be transmitted in both directions between the same two points simultaneously. Such systems are generally four-wire systems—a signal and a reference line for information transmission in each direction. FDX operation can be achieved with two-wire lines through the use of frequency-division multiplexing. Two-wire FDX operation is explained in Chapter 7.

4. *Full/full duplex* (F/FDX):   In a F/FDX system, the primary has the capability of transmitting to one secondary while receiving from a different secondary simultaneously. This classification is restricted in use to a multipoint system.

These four terms—simplex, half-duplex, full duplex, and full/full duplex—are used not only to describe a system but also to describe the mode of operation. In the multipoint system of Figure 2–2, the primary and the secondaries all have four-wire connections. However, the primary is capable of operating in the F/FDX mode while the secondaries operate only in the HDX mode.

DATA COMMUNICATIONS

A simplified block diagram of a data communications link is shown in Figure 2–4. Although only one secondary is shown, it represents a typical secondary in a multipoint system. One of the functions of the host computer is to store the applications programs for the various secondaries. The secondaries may have similar or dissimilar functions. The same control code received from two different secondaries may warrant two different actions on the part of the host computer. If the system so requires, the host computer must also provide storage for the centralized data and the software for database management. Data information is normally stored in auxiliary memories to which the computer has ready access. The peripherals connected to the host computer allow

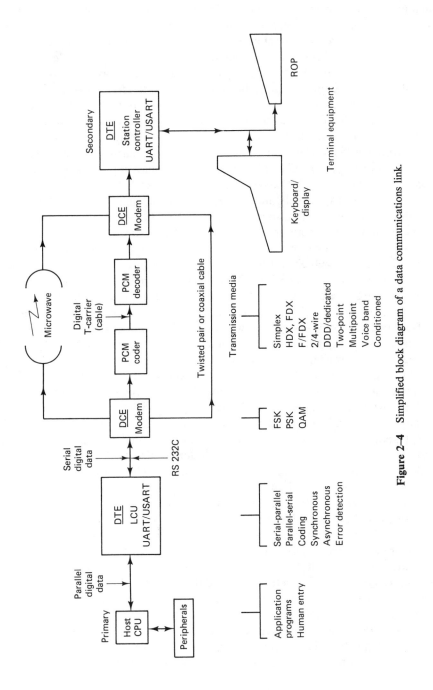

Figure 2–4 Simplified block diagram of a data communications link.

for human entry and intervention. As needs change, application programs may require additions, deletions, or changes. Data link problems also arise which are beyond the range of the computer's programmed capabilities. In such cases, human entry is required to make the necessary changes to resolve the existing problems.

The end equipment which either generates the digital information for transmission or uses the received digital data can be computers, printers, keyboards, CRTs, and so on. This equipment generally manipulates digital information internally in word units— the number of bits that make up a word in a particular piece of equipment are transferred in parallel. Digital data, when transmitted, are in serial form. Parallel transmission of an 8-bit word would require eight pairs of transmission lines—not at all cost-effective. *Data terminal equipment* (DTE) is a general phrase encompassing all of the circuitry necessary to perform parallel-to-serial and serial-to-parallel conversions for transmission and reception respectively and for data link management. The *UART* (Universal Asynchronous Receiver/Transmitter) and *USART* (Universal Synchronous/Asynchronous Receiver/Transmitter) are the devices that perform the parallel-to-serial and serial-to-parallel conversions (Chapter 5). The primary DTE includes a *line control unit* (LCU or LinCo) which controls the flow of information in a multipoint data link system. A *station controller* (STACO) is the corresponding unit at the secondaries. If there is software associated with the LCU, it is then called a *front-end processor* (FEP). At one time, the DTE was the last piece of equipment that belonged to the subscriber in a data link system. Between the DTEs, starting with the modems, was communications equipment owned and maintained by Telco. Recent judgments have removed modems from the realm of exclusive Telco property.

Data communications equipment (DCE) accepts the serial data stream from the DTE and converts it to some form of analog signal suitable for transmission on voice-grade lines. At the receive end, the DCE performs the reverse function of converting the received analog signal to a serial digital data stream. The simplest form of DCE is a *modem* (modulator/demodulator) or *data set.* At the transmit end, the modem can be considered a form of digital-to-analog converter, while at the receive end, it can be considered a form of analog-to-digital converter. The most common forms of modulation by modems are *frequency shift keying* (FSK), *phase shift keying* (PSK), and *quadrature amplitude modulation* (QAM) (Chapter 7).

If the cables and signal levels used to interconnect the DTE and DCE were left unregulated, the variations generated would probably be proportional to the number of manufacturers. Electronics Industries Association (EIA), an organization of manufacturers concerned with establishing industry standards, have agreed on the RS 232C as the standard interface between the DTE and the modem. This is a 25-pin cable whose pins have designated functions and specified signal levels (Chapter 6). The RS 232C is anticipated to be replaced by an updated standard.

All information that leaves the modem is in analog form suitable for transmission over a voice-grade line. The alternatives for the actual transmission of these signals, at this point, are identical to voice which occupies the same frequency range:

1. This information can be transmitted directly on twisted pairs or coaxial cables. The twisted pairs may be dial-up or dedicated lines.

2. This information could further amplitude-modulate other carrier frequencies to produce single sidebands. These single sidebands would be frequency-division multiplexed with other sidebands similarly derived and then transmitted on a microwave link (Chapter 8).

3. This information could be converted to a different form of digital signal through a process of pulse-code modulation (PCM) and then transmitted on T-carriers (Chapter 9).

Whichever the method, the reverse process must be accomplished at the receiver to reconstruct the original data stream.

For successful transfer of digital information over an analog link, three types of synchronization must be accomplished:

1. *Carrier synchronization* between the modems so that the received signal may be properly demodulated.

2. *Bit synchronization*, so that the receive DTE knows when to sample the incoming bit stream. Errors would be introduced if either a bit was sampled twice (sampling rate too fast) or not at all (sampling rate too slow).

3. *Character synchronization.* It is one thing to identify the value of each bit and another to associate it properly with the word to which it belongs. There must be some means of identifying where each word starts and ends.

   The transmitted data and the modems are each classified as either synchronous or asynchronous.

   (a) *Synchronous data* implies the inclusion of unique characters in the transmission for the purpose of establishing character synchronization. These characters are transmitted as the first part of any message.

   (b) *Asynchronous data* implies that character synchronization is achieved by framing each transmitted character with a start bit and stop bit (s).

   (c) *If a modem is synchronous*, it extracts clocking information from the received message and its own internal clock is synchronized to this signal.

   (d) *If the modems are operating asynchronously*, the internal clocks of the transmit and the receive modems are operating independently of one another.

Generally, synchronous data use synchronous modems and asynchronous data use asynchronous modems. Synchronous modems are used where higher bit transmission rates are required. It is unlikely that synchronous data would ever be transmitted using asynchronous modems. However, asynchronous data are sometimes transmitted using synchronous modems. Such transmissions are termed *isochronous.*

Thus far, we have been concerned with the hardware involved in communications and the types of electrical signals used. In data communications, several other aspects must be examined. The first is line protocol. *Line protocol* is a set of rules governing the transmission of digital information. In a data communications system, some of the aspects which these rules must clarify are:

1. How modem synchronization is to be achieved.
2. How the primary will select a particular secondary.
3. How the primary will tell the secondary that it is about to receive information.
4. What the secondary should do if it is not ready to receive this information.
5. What should be done if there is an error in transmission.
6. How the primary gives a particular secondary permission to transmit.
7. What form of error detection should be used. In addition, *message protocol* must be established. If the transmitted information is strictly text, message protocol is unnecessary. However, if the primary must supply message forms for completion by the secondary, agreement must be reached on:

   (a) What signals will be used to control cursor movement on the CRT.
   (b) How the secondary operator can be prevented from overwriting the transmitted form.
   (c) If the same form is to be used for many items, how only operator-entered information can be transmitted back to the primary and the basic form retained by the secondary for reuse (Chapters 10, 11, and 12).

QUESTIONS

1. What are the lines called on which digital data are transmitted directly?
2. What is the difference between a codec and a modem?
3. ASCII is a form of raw digital data. (T, F)
4. Modems are synonymous with data sets. (T, F)
5. In digital communications, the source information can be either digital or analog. (T, F)
6. In data communications, the source information can be either digital or analog. (T, F)
7. Contention can be a problem in a multipoint system. (T, F)
8. In a two-point communications system, one (the same) end is always identified as the master. (T, F)
9. Secondaries cannot communicate directly with each other in a multipoint system. (T, F)
10. Applications programs may be stored at either the primary or the secondary. (T, F)
11. A tributary is an alternate name for the primary. (T, F)
12. Explain the difference between simplex, HDX, FDX, and F/FDX.
13. What is an ROP?
14. What is the name of the interface between the DTE and the DCE?
15. Bits are transferred in parallel between the DTE and the DCE. (T, F)
16. The UART would be found in the DTE. (T, F)
17. The modem would be found in the DTE. (T, F)
18. What is meant by a "synchronous modem"?
19. What is meant by "asynchronous data"?

20. Digital source information could be converted to analog information and then back to digital information before transmission.   (T, F)

21. Voice source information is always in analog form from source to destination.   (T, F)

22. What are the three types of synchronization required in data communications?

23. Define line protocol.

24. Line protocol does not concern itself at all with the message contents.   (T, F)

25. Identify three types of modulation performed by modems.

3

THE TELEPHONE CIRCUIT

A *telephone circuit* consists of two or more facilities, interconnected in tandem, to provide a transmission path between a source and a destination. The interconnects may be temporary, as in a dial-up circuit, or permanent private-line circuits. The facilities may be cable pairs or carrier systems, and the information may be transferred on a coaxial, metallic, microwave, optic fiber, or satellite communications system. The information transferred is called the *message* and the circuit used is called the *message channel*. Telco offers a wide assortment of message channels ranging from a basic 4-kHz voice band circuit to wideband (30-MHz) microwave channels that are capable of transferring high-resolution video signals. The following discussion will be limited to a basic voice band circuit. In Telco terminology, the word *message* originally denoted speech information. This definition has been extended to include any standard voice frequency signal. Thus a message channel may include the transmission of speech, supervisory signaling, or voice band data.

THE LOCAL LOOP

The local loop is the only Telco facility required by all voice band data circuits. It is the primary cause of attenuation distortion and phase distortion. A local loop is a metallic transmission line (cable pair), consisting of two insulated conductors twisted together. The insulating material may be wood pulp or polyethylene plastic, while the wire conductor is usually copper and, in some instances, aluminum. Wire pairs are stranded together into units. Adjacent wire pairs within a unit are twisted with different pitch (twist length). This reduces the undesired effects of inductive coupling between pairs and helps to eliminate crosstalk. Units are cabled together into cores and then placed inside a plastic sheath. Depending on the insulating material used, sheaths contain between 6 and 900 pairs of wire. Sheaths are connected together and strung between distribution frames within Telco central offices and junction boxes located in manholes, back alleys, or telephone equipment rooms within large building complexes. The length of a subscriber loop depends on the station location relative to a local central telephone office.

The transmission characteristics of a cable pair depend on the wire diameter, conductor spacing, dielectric constant of the insulator, and the conductivity of the wire. These physical properties, in turn, determine the inductance, resistance, capacitance, and conductance of the line. The resistance and inductance are distributed along the length of the wire, while the conductance and capacitance exist between the two wires. If the insulation is good, the effect of conductance is generally negligible.

The electrical characteristics of a cable (Figure 3–1) are uniformly distributed along its length and are appropriately referred to as *distributed parameters.* Since it is cumbersome working with distributed parameters, it is common practice to lump them into discrete values per unit length (i.e., millihenrys per 1000 ft). The amount of attenuation and phase delay experienced by a signal propagating down a line is a function of the *frequency* of the signal and the *electrical characteristics* of the cable pair. Figure 3–2 illustrates the effect of frequency on attenuation for a given length of line. On this unloaded wire, a 3000-Hz signal suffers 6 dB more attenuation than a 500-Hz signal on the same line. The cable acts as a low-pass filter for the signal. Extensive studies of attenuation on cable pairs have shown that a reduction in attenuation can be achieved if inductors are added in series with the wire. This is called *loading.* Loaded cable is identified by the addition of the letters H, D, or B and an inductance value to the wire gauge number. H, D, and B indicate that the loading coils are separated by 6000, 4500, and 3000 ft, respectively. Generally, the amount of series inductance added is either 44, 88, or 135 mH. A cable pair with the designation 26H88 is made of 26-gauge wire with 88 mH of series inductance added every 6000 ft. If a loaded cable is used, a 3000-Hz signal will suffer only 1.5 dB more loss than a 500-Hz signal. Note that the loss versus frequency characteristics for a loaded cable are relatively flat up to approximately 2000 Hz.

The low-pass filter characteristics of a cable also affect the phase distortion versus frequency characteristics of a signal. The amount of *phase distortion* is proportional to the *length* and *gauge* of the wire. Loading a cable also affects the phase characteristics of a line. The telephone company must often add gain and delay equalizers to a circuit in order to achieve the minimum requirements. Equalizers introduce discontinuities or ripples in the bandpass characteristics of a circuit. Automatic equalizers in modems are sensitive to this condition and, very often, an

Figure 3–1   Electrical characteristics of a transmission line.

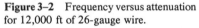

Figure 3–2 Frequency versus attenuation for 12,000 ft of 26-gauge wire.

overequalized circuit causes as many problems to a data signal as an underequalized circuit.

TRANSMISSION PARAMETERS

Transmission parameters are divided into three broad categories:

1. *Bandwidth parameters*, which include:
   (a) Attenuation distortion
   (b) Envelope delay distortion
2. *Interface parameters*, which include:
   (a) Terminal impedance
   (b) In-band and out-of-band signal power
   (c) Test signal power
   (d) Ground isolation
3. *Facility parameters*, which include:
   (a) Noise measurements
   (b) Frequency and phase distortion
   (c) Amplitude distortion
   (d) Nonlinear distortion

Bandwidth Parameters

The only transmission parameters with limits specified by the Federal Communications Commission (FCC) are attenuation distortion and envelope delay. *Attenuation distortion* is the difference in circuit gain experienced at a particular frequency with respect to the circuit gain of a reference frequency. This characteristic is also called *frequency response, differential gain,* and *1004-Hz deviation. Envelope delay* is an indirect method of evaluating the phase delay characteristics of a circuit. FCC tariff number 260 specifies the limits for attenuation distortion and envelope delay distortion. The limits are prescribed by line conditioning requirements. Through line conditioning, the attenuation and delay characteristics of a circuit are artificially altered to meet prescribed limits. Line conditioning is available only to private-line subscribers for an additional monthly charge. The basic voice band, 3002 channel satisfies the minimum line conditioning requirements.

Telco offers two types of line conditioning, C type and D type. *C-type conditioning* specifies the maximum limits for attenuation distortion and envelope delay distortion. *D-type conditioning* sets the minimum requirement for signal-to-noise ratio (S/N) and deals with nonlinear distortion.

D-type conditioning is referred to as high-performance conditioning and has two categories, D1 and D2. Limits imposed by D1 and D2 conditioning are identical. D1 conditioning is available for two-point circuits, while D2 conditioning is available for multipoint arrangements. D-type conditioned circuits must meet the following specifications:

1. Signal-to-C-notched noise: $\geq$ 28 dB
2. Nonlinear distortion:
   (a) Signal to second order: $\geq$ 35 dB
   (b) Signal to third order: $\geq$ 40 dB

D-type conditioning is mandatory when the data transmission rate is 9600 bps. If a facility is assigned by Telco for use as a 9600-bps circuit, and it does not meet the minimum requirements of D-type conditioning, it is so identified by Telco data technical support personnel and never considered for that purpose again. A different facility is sought. A circuit cannot be upgraded to meet D-type conditioning requirements by adding corrective devices.

C-type conditioning pertains to line impairments for which compensation can be made, to a certain degree, by filters and equalizers. This is accomplished with Telco-provided equipment. When a circuit is turned up for service with a particular C-type conditioning, it must meet the minimum requirements for that type of conditioning. The subscriber may include devices within the station equipment that compensate for minor long-term variations in the bandwidth requirements.

There are five classifications of C-type conditioning:

1. C1 and C2 pertain to two-point and multipoint circuits.
2. C4 pertains to two-point circuits and multipoint arrangements that have a maximum of four station locations.
3. C5 specifications pertain only to two-point circuits.
4. C3 conditioning is for access lines and trunk circuits associated with private switched networks.

Private switched networks are telephone systems dedicated to a single customer, usually with a large number of stations. An example is a large corporation with offices and complexes at different geographical locations with an on-premise private branch exchange (PBX) at each location. A PBX is a low-capacity dial switch where the subscribers are generally limited to stations within the same building complex. Common-usage access lines and trunk circuits are required to interconnect the PBXs. They are common only to the subscribers of the private network and not to the entire public telephone network.

Table 3–1 shows the various limits prescribed by the different types of C-

TABLE 3–1   BANDWIDTH PARAMETER LIMITS

Channel conditioning	Attenuation distortion (frequency response) relative to 1004 Hz		Envelope delay distortion	
	Frequency range (Hz)	Variation (dB)	Frequency range (Hz)	Variation (μs)
Basic	500–2500	+2 to −8	800–2600	1750
	300–3000	+3 to −12		
C1	1000–2400	+1 to −3	1000–2400	1000
	300–2700	+2 to −6	800–2600	1750
	300–3000	+3 to −12		
C2	500–2800	+1 to −3	1000–2600	500
	300–3000	+2 to −6	600–2600	1500
			500–2800	3000
C3 (access line)	500–2800	+.5 to −1.5	1000–2600	110
	300–3000	+.8 to −3	600–2600	300
			500–2800	650
C3 (trunk)	500–2800	+.5 to −1	1000–2600	80
	300–3000	−.8 to −2	600–2600	260
			500–2800	500
C4	500–3000	+2 to −3	1000–2600	300
	300–3200	+2 to −6	800–2800	500
			600–3000	1500
			500–3000	3000
C5	500–2800	+.5 to −1.5	1000–2600	100
	300–3000	+1 to −3	600–2600	300
			500–2800	600

conditioning on attenuation distortion and envelope delay. Figures 3–3 through 3–7 show a graphic presentation of several of the bandwidth parameter limits.

Attenuation distortion and C-type conditioning. The attenuation distortion limits for a basic 3002 channel require the circuit gain at any frequency between 500 and 2500 Hz to be not greater than 2 dB above the circuit gain at 1004 Hz and not more than 8 dB below the circuit gain at 1004 Hz (Figure 3–3). For attenuation distortion, the circuit gain at 1004 Hz is always used as the reference. Also, within the frequency bands from 300 to 499 Hz and from 2501 to 3000 Hz, the circuit gain cannot be greater than 3 dB above or more than 12 dB below the gain at 1004 Hz.

Example

A 1004-Hz test tone is transmitted at 0 dBm and received at −16 dBm. The circuit gain is − 16 dB (this is actually a loss of 16 dB: negative dB values are negative gains, or losses). Frequencies from 500 to 2500 Hz must be received at a minimum level of − 24 dBm and at a maximum level of − 14 dBm. Frequencies within the bands of 300 to 499 Hz and 2501 to 3000 Hz must be received at signal strength levels between − 13 and − 28 dBm inclusive. If the same transmission levels were applied to a circuit with C2 conditioning, frequencies between 500 and 2800 Hz must be received at signal levels ranging from − 15

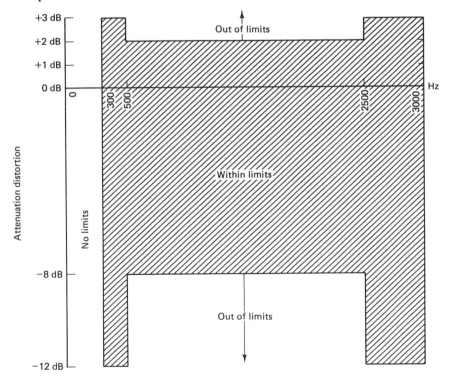

Figure 3–3 Graphic presentation of the limits for attenuation distortion in a basic 3002 channel.

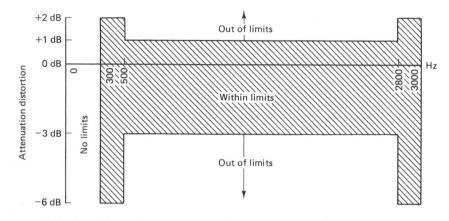

Figure 3–4   Graphic presentation of the limits for attenuation distortion on a channel with C2 conditioning.

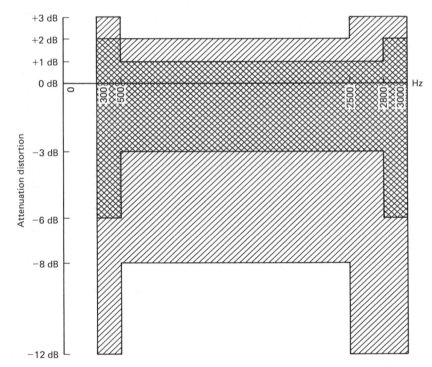

Figure 3–5   Overlay of Figures 3–3 and 3–4 to demonstrate the more stringent requirements imposed by C2 conditioning.

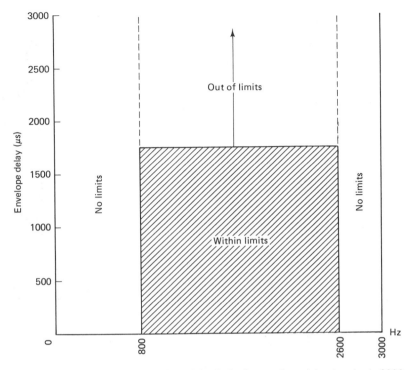

Figure 3–6 Graphic presentation of the limits for envelope delay in a basic 3002 channel.

to − 19 dBm. Frequencies within the bands of 300 to 499 Hz and 2801 to 3000 Hz must be received at signal levels ranging from − 14 to − 22 dBm. Table 3–1 shows that the higher the classification of conditioning imposed on a circuit, the flatter the frequency response, therefore, a better-quality circuit.

Envelope delay.   A linear phase versus frequency relationship is a requirement for error-free data transmission. This relationship is difficult to measure because of the difficulty in establishing a phase reference. Envelope delay is an alternate method of evaluating the phase versus frequency relationship of a circuit.

The time delay encountered by a signal as it propagates from source to destination is called *propagation time* or *phase delay*. All frequencies in the usable voice band (300 to 3000 Hz) do not experience the same time delay in a circuit. Therefore, a complex frequency spectrum, such as the output from a modem, does not possess the same phase versus frequency characteristics when it is received as when it was transmitted. This condition represents a possible impairment to a data signal. The *absolute phase delay* is the actual time required for a particular frequency to be

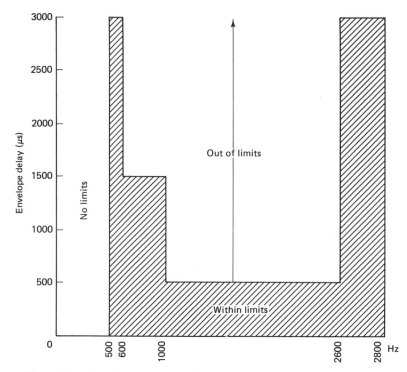

Figure 3–7   Graphic presentation of the limits for envelope delay on a channel with C2 conditoning.

propagated from source to destination. The difference between the absolute delays of different frequencies is phase distortion. A graph of phase delay versus frequency for a typical circuit is nonlinear.

By definition, envelope delay is the first derivative of phase with respect to frequency:

$$\text{envelope delay} = \frac{d\phi(\omega)}{d\omega}$$

In actuality, envelope delay only closely approximates $d\phi(\omega)/d\omega$. Envelope delay measurements do not evaluate true phase versus frequency characteristics, but rather, the phase of a wave that is the resultant of a narrow band of frequencies. It is a common misconception to confuse true phase distortion (also called delay distortion) with envelope delay distortion (EDD). *Envelope delay* is the time required to propagate a change in an AM envelope through a transmission medium. To measure envelope delay, a narrowband amplitude-modulated carrier, whose frequency is varied over the usable voice band, is transmitted. (The AM modulation rate is typically between 25 and 100 Hz.) At the receiver, phase variations of the low-frequency envelopes are measured. The phase difference at the different carrier frequencies is *envelope delay distortion*. The carrier frequency that produces the minimum envelope delay is established as the reference and is normalized to zero. Therefore, EDD measurements yield only positive values and indicate the relative envelope delays of the various carrier frequencies with respect to the reference frequency. The reference frequency of a typical voice band circuit is approximately 1800 Hz.

EDD measurements do not yield true phase delays, nor do they determine the relative relationships between true phase delays. EDD measurements are used to determine a close approximation of the relative phase delay characteristics of a circuit.

The EDD limit of a basic 3002 channel is 1750 μs between 800 and 2600 Hz, as shown on Table 3–1. This indicates that the maximum difference in envelope delay between any two carrier frequencies in this range cannot exceed 1750 μs.

Example

> An EDD test on a basic 3002 channel indicated that an 1800-Hz carrier experienced the minimum absolute delay of 400 μs. Therefore, it is the reference. The absolute envelope delay experienced by any frequency within the 800- to 2600-Hz band cannot exceed 2150 (400 + 1750) μs.

The absolute time delay encountered by a signal between any two points in the continental United States will never exceed 100 ms. This is not sufficient to cause any problems. Consequently, relative rather than absolute values of envelope delay are measured. For the previous example, as long as EDD tests yielded relative values less than + 1750 μs, the circuit is within limits.

Interface Parameters

The two primary considerations of the interface parameters are:

1. Electrical protection of the telephone network and its personnel
2. Standardization of design arrangements

These considerations are summarized below. Station equipment impedances should be 600 Ω resistive over the usable voice band, and the station equipment should be isolated from ground by a minimum of 20 MΩ dc and 50 kΩ ac. The basic voice-grade telephone circuit is a 3002 channel; it has a usable bandwidth of 300 to 3000 Hz. The circuit gain at 3000 Hz is 3 dB below the specified in-band signal power. The gain at 4 kHz must be at least 15 dB below the gain at 3 kHz. The maximum transmitted signal power for a private-line circuit is 0 dBm. The transmitted signal power for dial-up circuits is established for each loop so that the signal is received at the central Telco office at − 12 dBm.

Facility Parameters

Facility parameters represent potential impairments to a data signal. These impairments are caused by Telco equipment and the limits specified pertain to all voice band circuits regardless of conditioning.

1004-Hz variation.   Telco has established 1004 Hz as the standard test-tone frequency. The frequency 1004 Hz was selected because of its relative location in the passband of a standard voice band circuit. The purpose of this test tone is to simulate the combined signal power of a standard voice band transmission. The 1004-Hz channel loss for a private-line circuit is 16 dB. A 1004-Hz test tone applied to the transmit local loop at 0 dBm should appear at the output of the destination loop at -16 dBm. Long-term variations in the gain of the transmission facilities should not exceed ± 4 dB; the received signal power must be within the limits of -12 to -20 dBm.

Noise.   *Noise* can be generally defined as any undesired energy present in the usable passband of a communications channel. The noise is either correlated or uncorrelated. *Correlation* implies a relationship between the signal and the noise. Uncorrelated noise is energy present in the absence of a signal such as thermal noise. Correlated noise is unwanted energy which is present as a direct result of the signal, such as nonlinear distortion.

Noise weighting.   Signal interference by noise is categorized in terms of annoyance or intelligibility. Noise may be annoying to the listener but not to the degree that the conversation cannot be understood. Western Electric Company conducted experiments in which groups of listeners were asked to rate the annoyance caused by 14 different audible frequencies between 180 and 3500 Hz. These frequencies were presented to the listeners on a standard 500-type telephone (the old black dialer that Grandma had). The listeners first compared the annoyance of each frequency to the annoyance of a reference frequency of 1000 Hz in the absence of speech power. Then the same experiments were repeated with speech present. The results of the two tests were averaged and smoothed to produce the *C-message weighting curve.* This curve is shown in Figure 3–8.

C-message noise is significant because it indicates the passband characteristics that should be considered when conducting noise tests on voice band communications circuits. The frequencies from 600 to 3000 Hz proved to be the most annoying and, as the frequency decreased below 600 Hz, the annoyance factor also decreased. Because of these results, C-message filters were developed that have a response similar to the C-message weighting curve. These filters increase the attenuation of the noise power at frequencies below 600 Hz and above 3000 Hz, but they have a relatively flat response to noise power within the 600- to 3000-Hz passband. This response makes the measured strength of the noise frequencies proportional to the amount of annoyance they produce. These filters are inserted in noise-measuring sets at a point just before where the noise power is measured. The noise-measuring set then evaluates the noise in a manner similar to the human ear. The human ear cannot appreciate the true rms power of sound if the duration of the sound is 200 ms or less. Therefore, a noise-measuring set for voice circuits includes a 200-ms time constant that prevents them from reacting to short bursts of noise power. What significance does C-message weighting have on data circuits? Typical voice band data modems exhibit an output frequency spectrum that concentrates most of the transmitted signal power in the 600 to

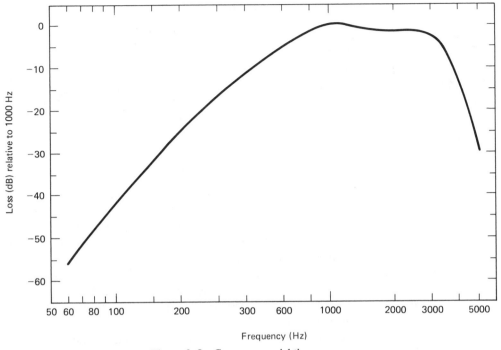

Figure 3–8 C-message weighting curve.

3000-Hz band. Since a C-message filter has a relatively flat frequency response over this range, noise measurements with this type of filter are valid for data applications also.

C-message noise. C-message noise measurements determine the average continuous rms noise power. This noise is commonly referred to as background, white, thermal, or Gaussian noise. This noise is inherently present in a circuit due to the electrical makeup of the circuit. Since white noise is additive, its magnitude is dependent, in part, on the electrical length of the circuit. C-message noise measurements are the terminated rms power readings at the receive end of a circuit with the transmit end terminated in the characteristic impedance of the telephone line (see Figure 3–9). There is a disadvantage in measuring noise this way. The overall circuit characteristics, in the absence of a signal, are not necessarily the same as when a signal is present. The use of companders and automatic gain devices in the circuit causes this difference. Companders are devices that exhibit frequency-dependent gain characteristics (companders are explained in detail in Chapter 9). For this reason C-notched noise measurements were developed.

C-notched noise measurements differ from standard C-message noise measurements only in the fact that a holding tone (usually, 1004 or 2804 Hz) is applied to the transmit end of the circuit during the test. The holding tone ensures that the circuit

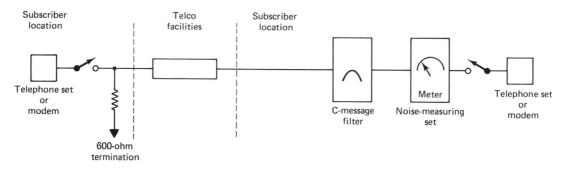

Figure 3–9   Terminated C-message noise test.

operation simulates a loaded voice or data transmission. "Loaded" is a communications term that indicates the presence of a signal power comparable to the power of an actual message transmission. The holding tone is filtered (notched out) in the noise-measuring set prior to the C-message filter. The bandpass characteristics of the notch filter are such that only the holding tone is removed. The noise power in the usable passband is measured with a standard C-message noise-measuring set. This test (Figure 3–10) assures that the noise readings obtained actually reflect the loaded circuit characteristics of a normal voice band transmission.

The physical makeup of a private-line data circuit may require the use of several trunk circuits in tandem. Each individual trunk may be an analog, digital, companded, or a noncompanded facility. Various combinations of these facilities may be used to configure a circuit. Telco has established realistic C-notched noise requirements for each type of facility for various trunk lengths. These requirements assist Telco in evaluating the performance of a facility and in expediting trouble-isolation procedures. A subscriber to the telephone network need only be concerned with the overall (end-to-end) C-notched noise requirement. Standard private-line data circuits, operating at less than 9600 bps, require a minimum signal-to-C-notched noise ratio of 24 dB. Data circuits operating at 9600 bps require a high-performance line with a minimum signal-to-C-notched noise ratio of 28 dB.

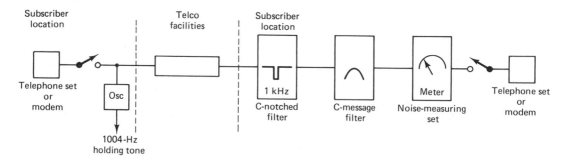

Figure 3–10   C-notched noise test.

Impulse noise.   *Impulse noise* is characterized by high-amplitude peaks (impulses) of short duration in the total noise spectrum. The significance of impulse noise hits on data transmission has been a controversial topic. Telco has accepted the fact that the absolute magnitude of the impulse hit is not as important as the magnitude of the hit relative to the signal amplitude. Empirically, it has been determined that an impulse hit will not produce transmission errors in a data signal unless it comes within 6 dB of the signal level (see Figure 3–11). Hit counters are designed to register a maximum of seven counts per second. This produces a "dead" time of 143 ms between counts when additional impulse hits are not registered. Contemporary high-speed data formats transfer data in block form and whether one hit or many occur during a single block is unimportant. Any error within a block necessitates retransmission of the entire block. Counting additional impulses during the time of a single block does not correlate well with data transmission performance.

Impulse tests are performed by placing a 2804-Hz holding tone on the circuit to ensure loaded circuit characteristics. The counter records the number of hits in a prescribed time interval (usually 15 minutes). An impulse hit is typically less than 4 ms and never more than 10 ms in duration. Telco's limit for recordable impulse hits is 15 within a 15-minute time interval. This does not limit the number of hits to one per minute but rather limits the average to one per minute.

Figure 3–11   C-notched and impulse noise.

Gain hits and dropouts.   A *gain hit* is a sudden, random change in the gain of a circuit. Gain hits are classified as temporary variations in the gain exceeding ± 3 dB, lasting more than 4 ms and returning to the original value within 200 ms. The primary cause of a gain hit is that of transients caused by switching radio facilities in the normal course of a day. Atmospheric fades produce the necessity for switching radio facilities. A *dropout* is a decrease in circuit gain of more than 12 dB that lasts longer than 4 ms. Dropouts are characteristics of a temporary open-circuit condition and are caused by deep radio fades or Telco maintenance activities. Dropouts occur at a rate of approximately one per hour. Gain hits and dropouts (see Figure 3–12) are detected by monitoring the receive level of a 1004-Hz test tone.

Phase hits.   *Phase hits* (see Figure 3–13) are sudden, random changes in the phase of a transmitted signal. Phase hits are classified as temporary variations in the phase of a signal that last longer than 4 ms. Generally, phase hits are not recorded unless they exceed $\pm 20°$ peak. Phase hits, like gain hits, are caused by transients produced when radio facilities are switched.

Phase jitter.   *Phase jitter* (Figure 3–13) is a form of incidental phase modulation—a continuous, uncontrolled variation in the zero crossings of a signal. Generally, phase jitter occurs at less than a 300-Hz rate, and its primary cause is a low-frequency ac ripple in Telco power plant supplies. The number of power supplies required in a circuit is directly proportional to the number of trunk circuits and Telco offices that make up a message channel. Each trunk facility has a separate phase jitter specification. The maximum end-to-end phase jitter allowed is $10°$ peak-to-peak regardless of the number of radio links, cable facilities, or digital carrier spans used in the circuit. Limiting the number of trunk circuits is a primary design consideration for a data circuit. Phase jitter is measured by observing the zero crossings of a 1004-Hz test tone.

Single-frequency interference.   *Single-frequency interference* is the presence of one or more continuous, unwanted tones within a message channel. The tones

Figure 3–12   Gain hits and dropouts.

Phase jitter
continuous changes < 10 p-p

Phase hit
rapid changes > 20 p

Figure 3–13   Phase distortion.

are called spurious tones and are often caused by crosstalk or cross-modulation between different channels in a carrier system. Spurious tones are measured by terminating the transmit end of a circuit, then searching through the channel spectrum with a frequency-selective voltmeter or observing the channel passband with a spectrum analyzer. Spurious tones can produce the same undesired circuit behavior as white noise.

Frequency shifts.   Analog carrier systems used by the telephone companies operate single-sideband suppressed carrier (SSBSC) and therefore require coherent demodulation. In *coherent demodulation,* the frequency of the suppressed carrier must be recovered and reproduced exactly by the receiver. If this is not done, the demodulated signal will be offset in frequency by the difference between the transmit and the receive carrier frequencies. Frequency shift is measured by transmitting a 1004-Hz test tone and then measuring the frequency of the tone at the receiver.

Phase intercept distortion.   *Phase intercept distortion* occurs in coherent SSBSC systems when the received carrier is not reinserted with the exact phase relationship to the received signal as the transmit carrier possessed. This impairment causes a constant phase shift to all frequencies. This impairment is of little concern with data modems that use frequency shift keying, differential phase shift keying, or quadrature amplitude modulation. Since these are the more common methods of modulation, the telephone company has not set any limits on phase intercept distortion.

Nonlinear distortion.   Nonlinear distortion is an example of correlated noise. The noise caused by nonlinear distortion is in the form of additional tones present because of the nonlinear amplification of a signal. No signal, no noise! Nonlinear distortion produces distorted sine waves. Two classifications of nonlinear distortion are:

1. *Harmonic distortion:*   unwanted multiples of the transmitted frequencies
2. *Intermodulation distortion:*   cross-products (sums and differences) of the input frequencies

Harmonic and intermodulation distortion, if of sufficient amplitude, can cause a data signal to be destroyed. The degree of circuit nonlinearity can be measured using either harmonic or intermodulation distortion tests.

Harmonic distortion is measured by applying a single-frequency test tone to a data channel. At the receive end, the power of the fundamental, second, and third harmonic frequences is measured. Harmonic distortion is classified as second, third, nth order or as total harmonic distortion. Generally, harmonics above the third extend beyond the passband of a voice band channel and are of insufficient amplitude to be important. The actual amount of nonlinearity in a circuit is determined by comparing the power of the fundamental to the combined powers of the second and third harmonics. Harmonic distortion tests use a single-frequency (704-Hz) source (see Figure 3–14); therefore, no cross-product frequencies are produced.

Although simple harmonic distortion tests provide an accurate measurement of the nonlinear characteristics of an analog message channel, they are inadequate for digital (T-carrier) facilities. For this reason, a more refined method was developed which uses a multifrequency test tone signal. Four test frequencies are used (see Figure 3–15): two designated the A band (A1 $=$ 856 Hz, A2 $=$ 863 Hz), and two designated the B band (B1 $=$ 1374 Hz, B2 $=$ 1385 Hz). The four frequencies are transmitted with equal power, and the total combined power is equal to that of a composite data signal. The nonlinear amplification of the circuit produces multiples of each frequency and also their cross-products. For reasons beyond the scope of this text, the following second- and third-order products were selected for measurement: B $+$ A, B $-$ A, and 2B $-$ A. The combined signal power of the four A and B band frequencies is compared with the second-order products, and then compared with the third-order products. Harmonic and intermodulation distortion tests do not directly determine the amount of interference caused by nonlinear circuit gain. They serve as a figure of merit only when evaluating circuit parameters.

Peak-to-average ratio.   The difficulties encountered in measuring true phase distortion or envelope delay distortion led to the development of peak-to-average (PAR) tests. A signal containing a series of distinctly shaped pulses with a high peak-to-average ratio (hence, the name) is transmitted. Delay distortion in a circuit has a tendency to spread the pulses and reduce the peak-to-average ratio. The received peak-to-average ratio is converted to a number between 0 and 100. The higher the number, the less the phase distortion. Peak-to-average tests do not indicate the exact amount of phase distortion present in a circuit; they only provide a figure of merit that can indicate the presence of a problem. Peak-to-average tests are less sensitive to attenuation distortion than envelope delay distortion tests and are easier to accomplish.

Figure 3–14   Harmonic distortion.

Figure 3–15 Nonlinear distortion.

SUMMARY

Tables 3–2 and 3–3, summarize the interface parameter limits and facility parameter limits discussed in this chapter.

TABLE 3–2   SUMMARY OF INTERFACE PARAMETER LIMITS[a]

Parameter	Limit
1. Recommended impedance of terminal equipment	600 Ω resistive ± 10%
2. Recommended isolation to ground of terminal equipment	At least 20 MΩ dc At least 50 kΩ ac At least 1500 V rms breakdown voltage at 60 Hz
3. Data transmit signal power	0 dBm (3-s average)
4. In-band transmitted signal power	2450- to 2750-Hz band should not exceed signal power in 800- to 2450-Hz band
5. Out-of-band transmitted signal power:	
Above voice band:	
(a) 3995–4005 Hz	At least 18 dB below maximum allowed in-band signal power
(b) 4–10 kHz band	Less than −16 dBm
(c) 10–25 kHz band	Less than −24 dBm
(d) 25–40 kHz band	Less than −36 dBm
(e) Above 40 kHz	Less than −50 dBM

Below voice band:

(f) Rms current per conductor as specified by Telco, but never greater than 0.35 A.

(g) Magnitude of peak conductor-to-ground voltage not to exceed 70 V.

(h) Conductor-to-conductor voltage shall be such that conductor-to-ground voltage is not exceeded. For an ungrounded signal source, the conductor-to-conductor limit is the same as the conductor-to-ground limit.

(i) Total weighted rms voltage in band from 50 to 300 Hz, not to exceed 100 V. Weighting factors for each frequency component (f) are: $f^2/10^4$ for f between 50 and 100 Hz, and $f^{3.3}/10^{6.6}$ for f between 101 and 300 Hz.

6. Maximum test signal power: same as transmitted data power

[a]Courtesy of AT&T

TABLE 3–3   SUMMARY OF FACILITY PARAMETER LIMITS[a]

Parameter	Limit
1. 1004-Hz loss variation	Not more than ±4 dB long-term
2. C-message noise	Maximum rms noise at modem receiver (nominal −16 dBm point)

Facility miles	*dBm*	*dBrncO*
0–50	−61	32
51–100	−59	34
101–400	−58	35
401–1000	−55	38
1001–1500	−54	39
1501–2500	−52	41
2501–4000	−50	43
4001–8000	−47	46
8001–16000	−44	49

Parameter	Limit
3. C-notched noise	(minimum values)
(a) Standard voice band channel	24-dB signal-to-C-notched noise
(b) High-performance line	28-dB signal-to-C-notched noise
4. Single-frequency interference	At least 3 dB below C-message noise limits
5. Impulse noise:	

Threshold with respect to 1004-Hz holding tone	*Maximum counts above threshold allowed in 15 minutes*
0 dB	15
+4 dB	9
+8 dB	5

Parameter	Limit
6. Frequency shift	±5 Hz end to end
7. Phase intercept distortion	No limits
8. Phase jitter	No more than 10° peak to peak (end-to-end requirement)
9. Nonlinear distortion (D-conditioned circuits only)	
Signal-to-second order	At least 35 dB
Signal-to-third order	At least 40 dB
10. Peak-average ratio	Reading of 50 minimum end to end with standard PAR meter
11. Phase hits	8 or less in any 15-minute period greater than ±20° peak
12. Gain hits	8 or less in any 15-minute period greater than ±3 dB
13. Dropouts	2 or less in any 15-minute period greater than 12 dB

[a]Courtesy of AT&T.

Abbreviations Associated with Signal and Noise Level Measurements

1. *dB (decibel):* Experiments indicate that a listener cannot give a reliable estimate of the loudness of a sound, but he can distinguish the difference in loudness between two sounds. The ear's sensitivity to a change in sound power follows a logarithmic rather than linear scale, and the dB has become the unit of this change.

$$dB = 20 \log \frac{V_1}{V_2} \qquad dB = 20 \log \frac{I_1}{I_2} \qquad dB = 10 \log \frac{P_1}{P_2}$$

   If the larger value is assigned to the numerator, the dB value will be positive. dB yields a relative value—the relative size of the numerator compared to the denominator.

2. *dBm:*

   (a) dBm is the dB in reference to 1 mW.

   (b) dBm is a measure of absolute power.

   (c) A 10 dBm signal is equal to 10 mW.

   (d) A 20 dBm signal is equal to 100 mW.

3. *TLP (transmission level point):* Knowing the signal power at any point in a system is relatively worthless. A signal at a particular point can be 10 dBm. Is this good or bad? This could be answered if it is known what the signal strength should be at that point. The TLP does just that. The reference for TLP is 0 dBm. A -15 dBm TLP indicates that, at this specified point, the signal should be -15 dBm. A 0 TLP is a TLP where the signal power should be 0 dBm. The TLP says nothing about the signal itself.

4. *dBmO:* dBmO is a power measurement adjusted to 0 dBm that indicates what the power would be if it were measured at 0 TLP. This value compares the actual signal at a point to what that signal should be at that point (TLP). A $+4$ dBm signal measured at a -16 dBm TLP is $+20$ dBmO. It is 20 dB stronger than the reference. -2 dBm at a -10 dBm TLP is equal to $+8$ dBmO.

5. *rn (reference noise):* This value is the dB value used as noise reference. This value is *always* -90 dBm or 1 pW. This value is taken as the reference because the signal value at any point should never be less than this.

6. *dBrnc:* This is the dB value of noise with respect to reference noise with C-message weighting. The larger this value is, the worse the condition because of a larger amount of noise.

7. *dBrncO:* This is the amount of noise corrected to 0 TLP. The amount of noise at a point does not really indicate how detrimental its effect is on a given signal. 0.3 dBm of noise with a 20 dBm signal is insignificant, while the same amount of noise with a 0.3 dBm signal strength may completely obscure the signal. Noise of 34 dBrnc at $+7$ dBm TLP yields a value of 27 dBrncO. dBrncO relates noise

power readings (dBrnc) to 0 TLP. This unit establishes a common reference point throughout the system.

8. *DLP (data level point):*   This parameter is equivalent to the TLP as far as its function is concerned. It is used as a reference for data transmission. TLP is used as a reference for voice transmission. Whatever the TLP is, the DLP is always 13 dB below that value for the same point. If the TLP is -12 dBm, the DLP is -25 dBm for that same point.

Example:

Noise $=-72$ dBm (noise, in all cases, will mean C-message noise).
Signal $=-27$ dBm at -24 dBm TLP.
Find:

Signal _____ dBmO	answer -3
Noise _____ dBrnc	answer $+18$
Noise _____ dBrncO	answer $+42$
S/N _____ dB	answer $+45$

9. *dBrn:*   This value is used only when there is excessive low-frequency noise and that noise is exceeding the levels specified in the interface parameters. Its value is measured with a device that has a 3-kHz flat filter (instead of a C-message filter) (see Figure 3–16). In normal circuits, dBrn is typically 1.5 dB above what would be read with a C-message filter.

QUESTIONS

1. Higher frequencies generally suffer greater attenuation than lower frequencies when traveling through a given cable of a fixed length.   (T, F)

2. A loaded telephone cable identification number indicates three characteristics of that cable. What are they?

Figure 3-16 3k Hz Flat filter.

3. A metallic cable acts as what type of filter?

4. A cable is loaded by adding _____ in series with the wire.

5. 1004-Hz deviation is another name for envelope delay. (T, F)

6. Dial-up lines may be conditioned. (T, F)

7. What type of line has minimum conditioning?

8. D-type conditioning may be achieved by adding inductors in series to the line. (T, F)

9. D-type conditioning is mandatory if the transmission rate is _____ bps.

10. Which frequency is used as the reference for attenuation distortion measurements when determining the type of conditioning?

11. Which frequency is used as the reference for envelope delay measurements when determining the type of conditioning?

12. All frequencies take the same time to propagate down a particular cable of fixed length. (T, F)

13. In determining the amount of phase distortion, (*relative, absolute*) phase delay is the more important parameter.

14. A C4 conditioned line must meet (*more, less*) stringent requirements than a C2 conditioned line.

15. Phase distortion is caused by the fact that it takes different frequencies different times to propagate down a line. (T, F)

16. Phase distortion is synonymous with delay distortion. (T, F)

17. Delay distortion is synonymous with EDD. (T, F)

18. The maximum allowed transmitted signal power for a dial-up circuit is always 2 mW.   (T, F)

19. The maximum allowed transmitted signal power for a private-line circuit is _____ mW.

20. All voice band frequencies suffer the same attenuation when passing through a C-message filter.   (T, F)

21. C-message filters are used in circuits to improve the signal-to-noise ratio.   (T, F)

22. What is the difference in measuring C-message noise and C-notched noise?

23. For a private line, what is the minimum signal-to-C-notched noise for:
   (a) Transmission at less than 9600 bps?
   (b) 9600-bps transmission?

24. Empirical determinations are the results of mathematical proofs of theory.   (T, F)

25. Three impulse hits on a message block would necessitate a different action than a single impulse hit on the same block.   (T, F)

26. What is the difference between a gain hit and an impulse hit?

27. Name two types of nonlinear distortion.

28. dB is a (*relative, absolute*) value.

29. dBm is a (*relative, absolute*) value.

30. What unit of measurement indicates the actual signal strength relative to what the signal strength should be?

31. What is the magnitude of noise, in dBm, that is used as the reference?

32. What is the unit of noise measurement that is corrected to 0 TLP?

33. Data transmission relates a signal strength to a DLP which is always _____ dB below the TLP?

34. What is the noise unit that is used to measure low-frequency noise?

PROBLEMS

1. A cable is identified as 16D44. This indicates that _____ -mH inductors have been added in series every _____ feet to a _____ -gauge wire.

2. A 1800-Hz signal takes 2 ms to propagate down a line. A 600-Hz signal takes 2.5 ms to travel down the same line. What is the absolute phase delay of the 600-Hz signal? What is the relative phase delay of this same signal?

3. A line has C4 conditioning. A 1004-Hz signal is received at -10 dBm and the propagation delay for a 1800-Hz signal is 5 ms. A signal frequency of 400 Hz can take no more than _____ ms to travel down this line and its received signal strength must be between _____ and _____ dBm.

4. Draw a graphic representation for attenuation distortion and envelope delay distortion for a channel with C4 conditioning.

5. Frequencies of 250 Hz and 1 kHz are applied to a C-message filter. Their difference in amplitude would be (*greater, the same, less*) at the output than at the input.

6. Dedicated lines are used to transmit data at 4800 bps. The received signal level is 500 μW. What is the maximum amount of C-notched noise allowed to satisfy D-type conditioning?

7. A 2000-Hz carrier is amplitude-modulated by a frequency of 300 Hz. Identify the type of nonlinear distortion if:

   (a) 1500 Hz is present at the output

   (b) 1700 Hz is present at the output

8. A C-message noise measurement taken at a -22 dBm TLP indicates -72 dBm of noise. A test tone is measured at the same TLP and measures -25 dBm. Determine the following signal/noise levels:

   (a) Signal power relative to TLP (dBmO)

   (b) C-message noise relative to reference noise (dBrnc)

   (c) C-message noise relative to reference noise at a 0 TLP (dBrncO)

   (d) Overall signal-to-noise ratio (dB)

9. A C-message noise measurement taken at -20 dBm TLP indicates a corrected reading of 43 dBrncO. A test tone transmitted at data level is used to determine a signal-to-noise of 31 dB. Determine the following signal/noise levels:

   (a) Signal power relative to TLP (dBmO)

   (b) C-message noise relative to reference noise (dBrnc)

   (c) Actual signal power of test tone (dBm)

   (d) Actual C-message noise power

10. A test-tone signal power of -62 dBm is measured at -61 dBm TLP. The C-message noise is measured at the same TLP and measures -10 dBrnc. Determine the following signal-to-noise levels:

   (a) C-message noise relative to reference noise at a 0 TLP (dBrncO)

   (b) Actual C-message noise power (dBm)

   (c) Signal power relative to TLP (dBmO)

   (d) Overall signal-to-noise ratio (dB)

4

TRANSMISSION CODES AND ERROR DETECTION AND CORRECTION

TRANSMISSION CODES

Information generated by a computer or a data terminal is in digital form. The information is either raw digital data or text (message). The alphanumeric characters and symbols (i.e., +, ", commas, etc.) which make up the message must be encoded for computer storage. The most common codes used for message encoding are ASCII (American Standard Code for Information Interchange) and EBCDIC (Extended Binary-Coded Decimal Interchange Code).

ASCII

ASCII is a 7-bit code and is shown in Table 4–1. For the ASCII codes 5EH and 5FH, many ASCII tables show UP ARROW and LEFT ARROW, respectively. Often, two hexadecimal characters are used to express this code with an assumed 0 for the most significant bit. The ASCII code for A expressed in hexadecimal is 41H. The trailing H signifies hexadecimal. ASCII characters can be divided into three groups:

1. **Data Link Control Characters**
   SOH   Start of Heading
   STX   Start of Text
   ETX   End of Text
   EOT   End of Transmission

TABLE 4–1 ASCII

Bits 3, 2, 1, 0	Hex	Bits 6, 5, 4								Hex
		000 / 0	001 / 1	010 / 2	011 / 3	100 / 4	101 / 5	110 / 6	111 / 7	
0000	0	NUL	DLE	SP	0	@	P	`	p	
0001	1	SOH	DC1	!	1	A	Q	a	q	
0010	2	STX	DC2	"	2	B	R	b	r	
0011	3	ETX	DC3	#	3	C	S	c	s	
0100	4	EOT	DC4	$	4	D	T	d	t	
0101	5	ENQ	NAK	%	5	E	U	e	u	
0110	6	ACK	SYN	&	6	F	V	f	v	
0111	7	BEL	ETB	'	7	G	W	g	w	
1000	8	BS	CAN	(	8	H	X	h	x	
1001	9	HT	EM	)	9	I	Y	i	y	
1010	A	LF	SUB	*	:	J	Z	j	z	
1011	B	VT	ESC	+	;	K	[	k	{	
1100	C	FF	FS	,	<	L	\	l	¦	
1101	D	CR	GS	-	=	M	]	m	}	
1110	E	SO	RS	.	>	N	∧	n	∿	
1111	F	SI	US	/	?	O	—	o	DEL	

ENQ	Enquiry
ACK	Positive Acknowledgment
DLE	Data Link Escape
SYN	Synchronous Idle
NAK	Negative Acknowledgment
ETB	End of Transmission Block

The specific meaning and functional use of the link control characters are explained and demonstrated in Chapters 10 and 11.

2. **Graphic Control Characters**

BEL	Bell	DC3	Device Control 3
BS	Back Space	DC4	Device Control 4
HT	Horizontal Tab	EM	End of Medium
LF	Line Feed	ESC	Escape
VT	Vertical Tab	FS	File Separator
FF	Form Feed	GS	Group Separator
CR	Carriage Return	RS	Record Separator
SO	Shift Out	US	Unit Separator
SI	Shift In	DEL	Delete
DC1	Device Control 1	CAN	Cancel
DC2	Device Control 2		

These characters are transmitted to identify where information will be printed or displayed or to signal the operator.

3. **Alphanumeric Characters**

These are the remaining characters that were not included in the first two groups.

They are the numbers, letters, punctuation marks, and so on, that make up the message that is displayed on the CRT or printed.

EBCDIC

EBCDIC is an 8-bit code devised by IBM. This code is shown in Table 4–2. EBCDIC characters can be divided into the same three major categories as ASCII. The following are abbreviations that differ from ASCII:

BYP	Bypass	PF	Punch Off
CC	Unit Backspace	PN	Punch On
DS	Digit Select	PRE	Prefix
FF	Form Feed	RES	Restore
IFS	Interchange File Separator	SI	Shift In
IGS	Interchange Group Separator	SM	Start Message
IL	Idle	SMM	Repeat
IRS	Interchange Record Separator	SO	Shift Out
IUS	Interchange Unit Separator	SOS	Start of Significance
LC	Lower Case	SUB	Substitute
NL	New Line	UC	Upper Case
NUL	Null		

Caution should be used in using EBCDIC since IBM identifies the least significant bit as bit 7 and the most significant bit as bit 0.

Example

EBCDIC for $n = 10010101 = 95$H.

ERROR DETECTION

Parity

The most common method used for error detection is through the use of parity. A single bit called the *parity bit* is added to every word so that the total number of 1's in a word, together with the parity bit, is odd for odd parity and even for even parity.

Example

Assume that odd parity is desired and that ASCII is the code used. The code for A is P100 0001. P represents the parity bit. Since odd parity is desired, P should take on a value to make the total number of 1's odd. Therefore, P takes on a value of 1 and 1100 0001 would be transmitted.

The parity bit is inserted at the transmitter and parity is checked at the receiver. The receiver should know the type of parity used. If odd parity is expected, and a word,

TABLE 4–2 EBCDIC

Bit position 0 1 2 3 \ 4 5 6 7	0 0 0 0	0 0 0 1	0 0 1 0	0 0 1 1	0 1 0 0	0 1 0 1	0 1 1 0	0 1 1 1	1 0 0 0	1 0 0 1	1 0 1 0	1 0 1 1	1 1 0 0	1 1 0 1	1 1 1 0	1 1 1 1
0 0 0 0	NUL	SOH	STX	ETX	PF	HT	LC	DEL			SMM	VT	FF	CR	SO	SI
0 0 0 1	DLE	DC1	DC2	DC3	RES	NL	BS	IL			CC		IFS	IGS	IRS	IUS
0 0 1 0	DS	SOS	FS	SYN	BYP	LF	EOB	PRE			SM			ENQ	ACK	BEL
0 0 1 1			SYN		PN	RS	UC	EOT					DC4	NAK		SUB
0 1 0 0	SP										¢	.	<	(	+	|
0 1 0 1	&										!	$	*	)	;	¬
0 1 1 0	-	/											%	_	>	?
0 1 1 1											:	#	@	'	=	"
1 0 0 0		a	b	c	d	e	f	g	h	i						
1 0 0 1		j	k	l	m	n	o	p	q	r						
1 0 1 0			s	t	u	v	w	x	y	z						
1 0 1 1																
1 1 0 0		A	B	C	D	E	F	G	H	I						
1 1 0 1		J	K	L	M	N	O	P	Q	R						
1 1 1 0			S	T	U	V	W	X	Y	Z						
1 1 1 1		1	2	3	4	5	6	7	8	9						□

together with its parity bit, have an even number of 1's, at least a 1-bit error has occurred. A limitation with parity is that if there is a 2-bit error in a word, this error will not be detected. There are many circuits that can be used for parity generators and parity checkers. The least complex of these is shown in Figure 4–1.

If there are *n* bits in the word, $n - 1$ XOR gates must be used for this circuit to function as a parity generator. No matter how many bits there are in the word, the output of this circuit will always be a 1 if there is an odd number of 1's in the word; the output will always be 0 if there is an even number of 1's in the word. Therefore, if even parity is desired, the output of this circuit can be used directly for the parity bit. If odd parity is desired, the output of this circuit must be passed through an inverter to generate the proper parity bit. To be used as a parity checker, the word plus the parity bit are applied to the inputs of this circuit. If no errors are detected, the output will always be low for even parity and always high for odd parity. What is done if a parity error is detected is dependent on the system. In some systems, automatic request for retransmission is initiated on the detection of a single parity error. Sometimes the message can be understood despite the error. Some systems will not request retransmission but instead, display an unused symbol (⸮) for the character that had the parity error. This is called *symbol substitution*. If the received message is "Four sc⸮re and seven...," the missing character can be guessed. However, if the message is "Pay Joe $⸮00,000.00...," it would be wise to request retransmission.

Figure 4–1 Parity generator or parity checker.

Order of the Arrow

A problem arises in the interpretation of a string of 1's and 0's in serial transmission. The least significant bit is always transmitted first. There is no disagreement in identifying A as the first letter of ABC. However, if the ASCII codes for these letters were shown in the same sequence with a trailing parity bit:

P100 0001	P100 0010	P100 0011
b a	d c	f e

the transmission bit sequence would be incorrect. *a* to *b* would be transmitted first, then *c* to *d*, then *e* to *f*. These bits can be rearranged to show the proper transmission sequence as follows:

Case 1:   P100 0011   P100 0010   P100 0001

Now the rightmost bit can be identified as the first bit transmitted and it is also the least significant bit of the first character. Similarly, the leftmost bit is the last bit transmitted and is the parity bit of the last character. Case 1 demonstrates the order of the arrow to the right—the rightmost bit is transmitted first and is the least significant bit of the first character. If the order of the arrow is to the left, the leftmost bit is transmitted first and is the least significant bit of the first character. This is shown in case 2.

Case 2:   1000 001P   0100 001P   1100 001P
                             1   4   2   4   3   4

If the order of the arrow is to the left, to interpret the bit pattern correctly, the bits must be taken in groups of 8, reversed, and then interpreted. If this were accomplished, 41H, 42H, and 43H would be obtained, which are the correct ASCII codes for A, B, and C.

Datascopes are used to monitor data lines. The first bit received would appear farthest left on the CRT. This means that the bit pattern received would be displayed with the order of the arrow to the left. In order to interpret a datascope presentation, a bit pattern with the order of the arrow to the left must be understood.

Vertical Redundancy Check and Longitudinal Redundancy Check

If two parity bits are used, a single bit error can be detected and corrected. *VRC* (vertical redundancy check) is simply inserting and checking the parity bit for each word as previously described. A group of words is called a *block*. The specific number of words in a block is described more fully in Chapter 11. Additional parity bits are computed based on the first bit of every word, on the second bit of every word, and so on. These parity bits make up the *LRC* (longitudinal redundancy check).

Example

Odd parity is used for the VRC, even parity for the LRC.

	T	H	E	S	C	A	T	L R C
				P				
LSB	0	0	1	0	1	1	0	1
	0	0	0	0	1	0	0	1
	1	0	1	0	0	0	1	1
	0	1	0	0	0	0	0	1
	1	0	0	0	0	0	1	0
	0	0	0	1	0	0	0	1
MSB	1	1	1	0	1	1	1	0
VRC	0	1	0	0	0	1	0	*=0

* is computed as a VRC bit. The serial transmitted bit stream is shown on page 50 with the order of the arrow to the right.

01000011 00100000 01000101 11001000 01010100
   C   SP   E   H   T

00101111 01010100 11000001
  LRC   T   A

The bits making up the LRC are also called the *block check sequence.* This sequence can represent an ASCII character. In this case, 010 1111 is the ASCII code for /, which is called the *block check character* for this particular block.

For the example above, assume that bit 2 of the letter E was received in error as a 0. The VRC would detect a parity error for this character but, by itself, could not specify which bit was wrong. The LRC would also detect an error in the second bit of some word but, by itself, could not identify which word had the error. Together, however, they identify the bit at the intersection of bit 2 and the letter E as erroneous. An inversion of this bit will correct the error.

Cyclic Redundancy Check

The *CRC* (cyclic redundancy check) is normally used with transmissions using EBCDIC as the character code. Basically, a generating function $G(x)$ is "divided into" a data string of 1's and 0's, the quotient is discarded, and the remainder is transmitted as the CRC. The number of bits in the CRC is equal to the highest exponent of the generating function. The generating function used with EBCDIC is

$$G(x) = x^{16} + x^{12} + x^5 + 1 \qquad \text{where } x^0 = 1$$

This function is also referred to as CRC-16 (because of the high-order exponent) and is 16 bits or two bytes long. These two bytes of CRC are also called the *block check sequence* (BCS) or the *frame check sequence* (FCS). The corresponding EBCDIC characters are called the *block check characters* (BCCs) or the *frame check characters* (FCCs).

The exponents identify the bit positions containing a 1. Therefore, the $G(x)$ shown above could be written as

1 0 0 0 1 0 0 0 0 0 0 1 0 0 0 0 1
b_{16} b_{12} b_5 b_0

The "division" is not accomplished in the usual manner. In the subtract portion of the division, the remainder is obtained through an XOR operation rather than by an honest subtraction. Therefore, any $G(x)$ will "divide" into another number with an equal amount of digits one time regardless of whether $G(x)$ is larger or smaller.

Example

Assume the message

$$M(x) = x^7 + x^5 + x^4 + x^2 + x + 1$$

$$= 1 \ 0 \ 1 \ 1 \ 0 \ 1 \ 1 \ 1$$

The least significant bit is on the *left*.

For simplicity, a $G(x)$ of $x^5 + x^4 + x + 1$ will be used. $M(x)$ is first multiplied by the number of bits in the CRC [same as the highest exponent of $G(x)$], producing

$$x^5 (x^7 + x^5 + x^4 + x^2 + x + 1) = x^{12} + x^{10} + x^9 + x^7 + x^6 + x^5$$
$$= 1\ 0\ 1\ 1\ 0\ 1\ 1\ 1\ 0\ 0\ 0\ 0\ 0$$

which is now divided by $G(x)$.

```
                  (discarded)      1 1 0 1 0 1 1 1
        1 1 0 0 1 1 ) 1 0 1 1 0 1 1 1 0 0 0 0 0
                      1 1 0 0 1 1
                      ─────────
                        1 1 1 1 0 1
                        1 1 0 0 1 1
                        ─────────
                            1 1 1 0 1 0
                            1 1 0 0 1 1
                            ─────────
                              1 0 0 1 0 0
                              1 1 0 0 1 1
                              ─────────
                                1 0 1 1 1 0
                                1 1 0 0 1 1
                                ─────────
                                  1 1 1 0 1 0
                                  1 1 0 0 1 1
                                  ─────────
                                    0 1 0 0 1 = CRC
```

The CRC is appended to the message producing a transmitted bit stream of

$$1\ 0\ 1\ 1\ 0\ 1\ 1\ 1 \qquad 0\ 1\ 0\ 0\ 1$$

At the receiver, this stream is "divided" by the same $G(x)$ as used by the transmitter and if the remainder is zero, there was no error.

```
                              1 1 0 1 0 1 1 1
        1 1 0 0 1 1 ) 1 0 1 1 0 1 1 1 1 0 1 0 0 1
                      1 1 0 0 1 1
                      ─────────
                        1 1 1 1 0 1
                        1 1 0 0 1 1
                        ─────────
                            1 1 1 0 1 0
                            1 1 0 0 1 1
                            ─────────
                              1 0 0 1 1 0
                              1 1 0 0 1 1
                              ─────────
                                1 0 1 0 1 0
                                1 1 0 0 1 1
                                ─────────
                                  1 1 0 0 1 1
                                  1 1 0 0 1 1
                                  ─────────
                                    0 0 0 0 0 0      no error
```

The circuit used to generate the CRC is shown in Figure 4–2.

Operation of the CRC generating circuit.   Initially, all of the flip-flops (numbered boxes) of the shift register are cleared and the Compute BCC enable is high

Figure 4–2 CRC generating circuit.

for as long as the data stream is applied. During this time, data are outputted by way of AND gate 1.

When the last data bit has passed through, the Compute BCC enable is made low. The BCC characters are contained in the shift register. With Compute BCC enable low, the circuit acts as a 5-bit shift register and, on succeeding clock pulses, the BCS is outputted by way of AND gate 2.

Example

The flip-flops are identified by their number. The same $G(x)$ and data stream are used as in the preceding example.

At the transmitter:

Flip-flops:	1	2 3 4	5	Data in
Initial clear:	0	0 0 0	0	
				1
After shift 1:	1	1 0 0	1	
				0
After shift 2:	1	0 1 0	1	
				1
After shift 3:	0	1 0 1	0	
				1
After shift 4:	1	1 1 0	0	
				0
After shift 5:	0	1 1 1	0	
				1
After shift 6:	1	1 1 1	0	
				1
After shift 7:	1	0 1 1	0	
				1
After shift 8:	1	0 0 1	0	

CRC in the shift register.
Rightmost bit is transmitted first.

At the receiver:

Flip-flops:	1	2 3 4	5	Data in Incl. CRC
Initial clear:	0	0 0 0	0	
				1
After shift 1:	1	1 0 0	1	
				0
After shift 2:	1	0 1 0	1	
				1
After shift 3:	0	1 0 1	0	
				1
After shift 4:	1	1 1 0	0	
				0
After shift 5:	0	1 1 1	0	
				1
After shift 6:	1	1 1 1	0	
				1
After shift 7:	1	0 1 1	0	
				1
After shift 8:	1	0 0 1	0	
				0 (CRC)
After shift 9:	0	1 0 0	1	
				1
After shift 10:	0	0 1 0	0	
				0
After shift 11:	0	0 0 1	0	
				0
After shift 12:	0	0 0 0	1	
				1
After shift 13:	0	0 0 0	0	

All zeros. No error.

Variations

1. Some circuits have an inverter located at point X on the circuit diagram. This causes the 1's complement of the CRC to be transmitted. An inversion would also be required at the receiver.
2. Some circuits initially preset the flip-flops. At the receiver, after the CRC has been received, a constant character (not 0's) is produced.

Hamming Code

The Hamming code is one form of a forward error-correcting code (FEC). This means that the receiver has the capacity to detect and correct errors in the received transmission. The number of bits in the Hamming code is dependent on the number of bits in the data string. If m represents the number of bits in the data string and n

represents the number of bits in the Hamming code, then n is the smallest number such that

$$2^n \geq m + n + 1$$

The Hamming bits can be placed anywhere in the data string. Once these positions are decided on, they must remain fixed and must be known by the transmitter and the receiver.

Example

Assume that the data string consists of the 12 bits

1 0 1 1 0 0 0 1 0 0 1 0

An $n = 4$ is insufficient. An $n = 5$ must be used. An H is arbitrarily placed in the Hamming bit positions.

```
17 16 15 14 13 12 11 10 9  8  7  6  5  4  3  2  1
H  1  0  1  H  1  0  0  H  H  0  1  0  H  0  1  0
```

The bit positions must be identified as shown.

To determine the value of the Hamming bits, express bit positions as a 5-bit number ($n = 5$) and XOR all positions containing a 1.

```
  2      0 0 0 1 0
  6      0 0 1 1 0
         0 0 1 0 0
 12      0 1 1 0 0
         0 1 0 0 0
 14      0 1 1 1 0
         0 0 1 1 0
 16      1 0 0 0 0
         1 0 1 1 0    Hamming bits
```

The Hamming bits will replace the H's left to right, respectively.

The encoded transmitted data string is now

1 1 0 1 0 1 0 0 1 1 0 1 0 0 0 1 0

Assume that these data are received with an error in bit 14:

1 1 0 0 0 1 0 0 1 1 0 1 0 0 0 1 0

The Hamming bits are extracted and XORed with all bit positions containing a 1.

```
         1 0 1 1 0
  2      0 0 0 1 0
         1 0 1 0 0
```

```
 6    0 0 1 1 0
      ─────────
      1 0 0 1 0
12    0 1 1 0 0
      ─────────
      1 1 1 1 0
16    1 0 0 0 0
      ─────────
      0 1 1 1 0 = 14    position of error
```

If the error occurred in the Hamming bit or if there was more than a 1-bit error, this coding scheme, as described, is not able to detect and correct. Additional coding bits would be required.

Any scheme for error detection and correction involves the transmission of additional bits. The purpose is to eliminate the need of retransmission. But the inclusion of additional bits for error detecting and correcting increases the transmission time. Before any such scheme for FEC is implemented, the frequency of errors and the corresponding time for retransmission must be compared with the increased transmission time caused by the insertion of additional bits for FEC.

QUESTIONS

1. What is the purpose of the parity bits?
2. Parity is checked at the transmitter.   (T, F)
3. Name the three functional usages of the ASCII characters.
4. The LRC is made up of the individual parity bits for each word.   (T, F)
5. What bit is the most significant bit in EBCDIC?
6. If EBCDIC is the transmission code, what is used for error detection?

PROBLEMS

1. Given the following ASCII codes with the order of the arrow to the right. What are the corresponding hexadecimal values and the corresponding ASCII characters?
   (a) 1010111   **(b)** 0000010   **(c)** 0010110   **(d)** 0101111
2. Given the following ASCII codes with the order of the arrow to the left. What are the corresponding hexadecimal values and the corresponding ASCII characters?
   (a) 1101110   **(b)** 0101000   **(c)** 1011011   **(d)** 0010000
3. For Problem 1, if odd parity is used, what is the parity bit?
4. For Problem 2, if even parity is used, what is the parity bit?
5. **(a)** Compute the VRC and LRC for MARY. Use odd parity to compute the LRC and even parity to compute the VRC.
   (b) What ASCII character is represented by the LRC?
6. Compute the LRC (odd parity) and VRC (even parity) for your last name.
7. How many Hamming bits are required for each ASCII word?

8. **(a)** For the ASCII A (1000001), insert the required number of Hamming bits into every other location starting at the right:

$$1\ 0\ 0\ 0\ 0\ 0\ 1$$
$$X\ X\qquad \text{etc.}$$

(b) What are the Hamming bits used?

(c) If the fifth bit from the right was received in error, show how the Hamming code would identify the location of the error.

5

UNIVERSAL ASYNCHRONOUS RECEIVER / TRANSMITTER

Functionally, the UART is divided into two sections, the transmitter and receiver. It is *asynchronous* because the transmit section does not output any clocking information and the receive section does not require any clocking information from the received data. The main functions of the UART are:

1. To perform serial-to-parallel and parallel-to-serial conversion of data
2. To insert and check parity bits
3. To insert start and stop bits

Initially, the master reset (MR) is made active to set TSO, TEOC, and TBMT and to reset RDA, RPE, RFE, and ROR. Before the UART can be used for transmission or reception, certain information must be provided:

1. Will a parity bit be used?
2. If yes, will odd or even parity be used?
3. How many stop bits will be used?
4. How many bits will there be in each word (5 to 8)?

This information is required by both the transmit and receive section and is common to both. Appropriate signals are applied to the NPB, POE, NSB, and NDB1 and NDB2 inputs and then strobed (CS) into the control register.

TRANSMITTER SECTION

A simplified block diagram of the UART transmitter is shown in Figure 5–1. A high output on the Transmit Buffer Empty (TBMT) pin tells the computer that it can send data to the UART. The computer sends the appropriate number of data bits in parallel on TDO–TD7 and strobes them into the transmit buffer register by applying a low on TDS. At this time, TBMT goes low. If the transmit shift register is empty, the data from the transmit buffer register, together with the start and stop bits and the parity bit, are moved into this register from which they will be shifted out serially on the TSO pin at a rate equal to $\frac{1}{16}$ of the transmit clock frequency. The bit steering logic places the parity bit immediately after the last data bit. The minimum number of bits that can be transmitted for one word is 7 (1 start bit, 5 data bits, no parity bit, and 1 stop bit); the maximum number is 12 (1 start bit, 8 data bits, 1 parity bit, and 2 stop bits). When transmission starts, Transmit End of Character (TEOC) goes low. The sequence in which this information is transmitted is: start bit, LSB of data to MSB of data, parity bit (if present), and stop bit(s). After the last stop bit is transmitted, TEOC returns to a high, signaling that the next word can be loaded into the transmit shift register from the transmit buffer register. As soon as data are transferred from the transmit buffer register to the transmit shift register, TBMT again is set, requesting the next word from the computer. The use of these two registers in the transmit section is called *double buffering*. It permits the computer to find the next word and send it to the UART while the UART is serially outputting the information. The start bit is not optional and is always a logic 0 or space; the stop bit(s) are always logic 1's or marks.

RECEIVER SECTION

A block diagram of the receiver section is shown in Figure 5–2. The receive section has its own clock input, which is generally 16 times the incoming bit rate. When the receive section is not receiving any information, the Receive Serial Input (RSI) pin is held high. Since a high represents a logic 1, the receiver is said to be receiving *idle line 1's*. A high-to-low transition on this pin indicates the possibility of incoming data. The start bit verify circuitry waits for a period of eight clock cycles and then rechecks this input. If the signal has returned to a high, the transition is attributed to a noise spike and is ignored. However, if this input is still low, the receive section assumes that it has received a valid start bit. Eight clock cycles from the initial high-to-low transition places the receiver approximately at the center of the start bit. With this point as a reference, the input data stream is sampled every 16 clock cycles. The sampling, therefore, occurs at the midpoint of each received data bit. The incoming data bits are shifted serially into the receive shift register. After all of the data bits have been received (the receiver knows the number of data bits/word from the settings in the control register), the parity bit is checked. If there is a parity error, the Receive Parity Error (RPE) flag in the status register is set. After the parity bit, the receiver looks for the required number of stop bits. If the stop bits are not detected, the Receive Frame Error (RFE) flag in the status register is set. The Receive Data Available (RDA) flag of the status register is now

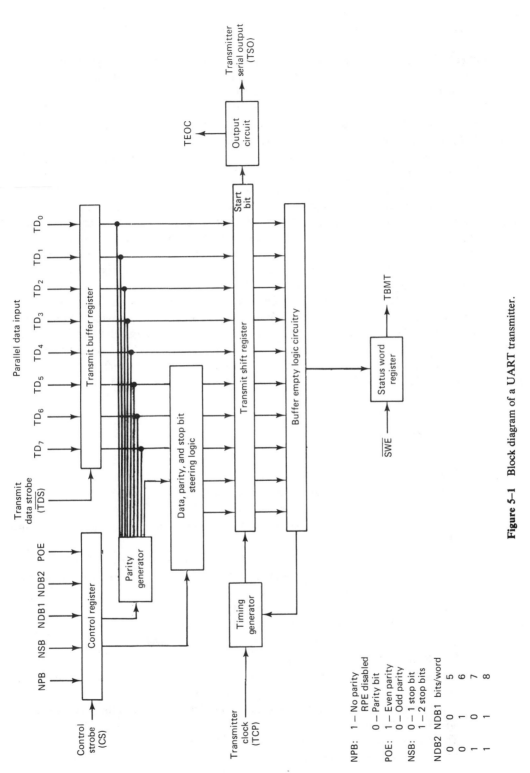

Figure 5-1 Block diagram of a UART transmitter.

NPB: 1 – No parity
                RPE disabled
       0 – Parity bit
POE: 1 – Even parity
       0 – Odd parity
NSB: 0 – 1 stop bit
       1 – 2 stop bits

NDB2	NDB1	bits/word
0	0	5
0	1	6
1	0	7
1	1	8

59

checked and, if reset, the data are moved in parallel from the receive shift register to the receive buffer register and the RDA flag is set. The RDA output indicates to the computer that there are data in the receive buffer register available for reading. To read the contents of the status register, the computer must activate the SWE input. Detecting RDA active, the computer reads the contents of the receive buffer register by activating RDE and then activating RDAR to reset the RDA flag.

Should the UART find the RDA flag set when a complete word has been shifted into the receive shift register, the Receive Overrun (ROR) flag of the status register is set after the word is transferred to the receive buffer register. Finding the RDA flag set indicates to the UART that the previous word has not yet been read by the computer. In the process of moving the current word from the receive shift register to the receive buffer register, the preceding word is destroyed. An active signal on ROR indicates to the computer that at least one word was lost because the computer did not read from the UART in time. The receive shift register and the receive buffer register provide double buffering in the receive section of the UART.

The stop bits were originally used to provide the receiving terminal with time to do something with the data received. With the speed of current devices, this is now a minimal problem. *Bit synchronization* at the receiver is achieved through the use of the receiver clock and the valid determination of the start bit. *Word synchronization* is achieved by knowing the number of bits per word (control register), and being able to detect that many bits between the start and stop bits.

QUESTIONS

1. List the primary functions of a UART.
2. A UART can operate in the half-duplex mode. (T, F)
3. A UART can operate in the full-duplex mode. (T, F)
4. A UART can transmit at one bit rate and receive at a different bit rate simultaneously. (T, F)
5. What signals must be applied to NDB2 and NDB1, respectively, to identify 7 bits per word?
6. What type of error is incurred if a second word is placed in the receive holding register before the first word is read out?
7. The status of RDA is always available on the RDA pin. (T, F)
8. If the computer sees an active signal on TBMT, what should it do next?
9. The parity bit is generated in the receive section of the UART. (T, F)
10. Which bit is not optional in a UART?
11. Why is the received data in a UART sampled at a rate much faster than the incoming bit rate?
12. If you wanted to increase the accuracy of the receive bit sampling, the receiver clock rate must be (*increased, decreased*).
13. The receive and transmit clocks in a UART must be synchronized. (T, F)
14. What is the minimum number of data bits that may be used with a UART?

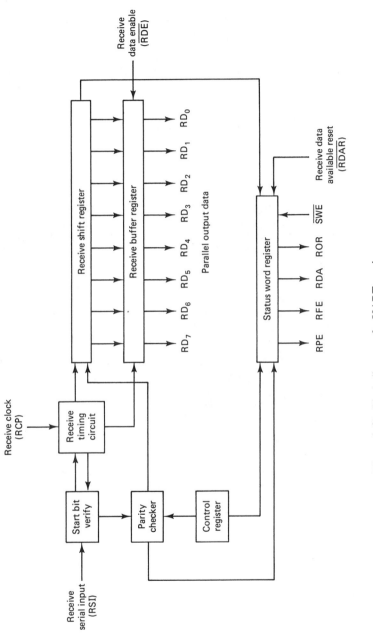

Figure 5–2 Block diagram of a UART receiver.

15. What is the purpose of the ROR status bit?

16. What is the maximum number of bits that may be transmitted for one character by a UART?

17. All UART transmissions contain a parity bit. (T, F)

18. A UART may be operated synchronously. (T, F)

19. The status bits must be determined before any data can be written into a UART. (T, F)

20. A UART is designed to transmit (*synchronous, asynchronous*) data.

PROBLEMS

1. The following ASCII transmission is received by a UART. Even parity is used with the order of the arrow to the left and 2 stop bits.

```
00010000111001000001110001100111101111011011011011011011
01101011111010010110110011101111101110011111000000010111
01110001011011110110110111101111100010011111101000010011
01100000011
```

(a) Translate the received bit pattern into English.

(b) Find the parity error.

2. The following ASCII transmission is received by a UART. Even parity is used with the order of the arrow to the left and two stop bits. Idle line 1's are present.

```
11111100100001011111110101010001111111110001100101111111
111111010101000111110100100111111111111111111110011000111 0
```

(a) Translate the message into English.

(b) Identify three errors in this transmission.

6

SERIAL INTERFACES

RS 232C INTERFACE

In an effort to standardize the interface between data terminal equipment and data communication equipment, Electronic Industries Association (EIA) agreed on the RS (recommended standard) 232C specifications. These specifications identify the electrical, mechanical, and functional description of this interface. The RS 232C is a 25-wire cable. The wires are used for the transmission of control signals or of digital data. Figure 6–1 identifies the voltages and terminology associated with the control and data lines. The drivers for these voltages are limited to a range of $+5$ to $+15$ V for a space or an ON condition: -5 to -15 V for a mark or an OFF condition. The difference between 3 and 5 V $(+$ or $-)$ is allowed for a noise margin. The driver voltages can be expected on-line provided that the line is terminated. If a line is not terminated, the line voltages should not exceed ± 25 V.

Data Lines

Logic 1	Logic 0
Mark	Space
Voltage of -3 V or more	Voltage of $+3$ V or more
Negative	Positive

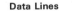
Control Lines

OFF	ON
Low	High
Voltage of -3 V or more	Voltage of $+3$ V or more
Negative	Positive

Figure 6–1 Data and control line voltages and terminology.

The maximum physical length is normally identified as 50 ft. The terminal capacitance is specified as 2500 pF, including cable capacitance. For the type of cable intended, the cable capacitance is approximately 50 pF per foot. From this, the 50-ft maximum cable length is derived. In most instances where a cable connection is identified as RS 232C compatible, the compatibility is restricted to the mechanical connectors and the line voltages specified by RS 232C. The maximum impedance at the terminating end is 7000 Ω. The functional use of each wire does not always conform to the function specified by EIA. EIA identifies each pin by a two- or three-letter designator. The functional use of each line is not immediately obvious from these letter designators.

RS 232C Interface*

Pin number	EIA nomenclature	Common or alternate nomenclature
1	Protective Ground (AA)	
2	Transmitted Data (BA)	Send Data (TD, SD)
3	Received Data (BB)	(RD)
4	Request to Send (CA)	(RS, RTS)
5	Clear to Send (CB)	(CS, CTS)
6	Data Set Ready (CC)	(DSR), Modem Ready (MR)
7	Signal Ground (AB) (Common Return)	
8	Received Line Signal Detector (CF)	(RLSD), Carrier on Detect (COD)
9	Reserved for Data Set Testing	$+V$ ($+10$ V dc)
10	Reserved for Data Set Testing	$-V$ (-10 V dc)
11	Unassigned	Equalizer Mode (QM)—Synchronous Modems
		Originate Mode (OM)—Asynchronous Modems
12	Secondary Received Line Signal Detector (SCF)	Local Mode (LM)—Asynchronous Modems
13	Secondary Clear to Send (SCB)	
14	Secondary Transmitted Data (SBA)	New Synch (NS)
15	Transmission Signal Element Timing (DB) (DCE Source)	Serial Clock Transmit (SCT)
16	Secondary Received Data (SBB)	Divided Clock Transmit (DCT)
17	Receiver Signal Element Timing (DD) (DCE Source)	Serial Clock Receive (SCR)
18	Unassigned	Divided Clock Receive (DCR)
19	Secondary Request to Send (SCA)	
20	Data Terminal Ready (CD)	(DTR)
21	Signal Quality Detector (CG)	(SQ)
22	Ring Indicator (CE)	(RI)

Pin number	EIA nomenclature	Common or alternate nomenclature
23	Data Signal Rate Selector (CH) (DTE/DCE Source)	(SS)
24	Transmit Signal Element Timing (DTE Source) (DA)	Serial Clock Transmit External (SCTE)
25	Unassigned	

*Courtesy Electronic Industries Association, EIA. This document, RS 232C, is available in its entirety from the EIA, Washington, D.C.

Signal Direction on RS 232C Interface

	2	Transmitted Data	→	
	3	Received Data	←	
	4	Request to Send	→	
	5	Clear to Send	←	
	6	Data Set Ready	←	
	8	Received Line Signal Detector	←	
D	12	Secondary Received Line Signal Detector	←	D
T	13	Secondary Clear to Send	←	C
E	14	Secondary Transmitted Data	→	E
	15	Transmission Signal Element Timing (DCE Source)	←	
	16	Secondary Received Data	←	
	17	Receiver Signal Element Timing (DCE Source)	←	
	19	Secondary Request to Send	→	
	20	Data Terminal Ready	→	
	21	Signal Quality Detector	←	
	22	Ring Indicator	←	
	23	Data Signal Rate Selector (DTE/DCE Source)	→	
	24	Transmit Signal Element Timing (DTE Source)	→	

Condensation of Typical Modem Specifications

Data Rate:	4800 bps
Modulation:	8-PSK
Operation:	Synchronous, binary, serial
Line Requirements:	Basic four-wire, private line (type 3002)
Operating Modes:	Simplex, half-duplex, duplex

Request to Send (RTS)/Clear to Send (CTS) delay:
  Switched carrier: 48.5 ms
  Continuous carrier with switched RTS option: 8 ms

Signal Quality Detector (SQ):
  Turn on if data carrier is present for 4 ms or longer.
  Turn off if data carrier is absent for 2 ms or longer.

Receive Line Signal Detect (RLSD):
  Turn on 41 ms after SQ on.
  Turn off less than 1 ms after SQ off. (1-second holdover option available.)

Protective Ground (1): Connected to the frame and to external ground at either the DCE or DTE but not both.

Signal Ground (7): Reference for control and data line voltages. Generally connected to protective ground to eliminate longitudinal power line noise.

Data Set Ready (6): Local data set is capable of transmitting and receiving data, and the data set is not in the TEST mode. An ON condition does not imply that a channel has been established to the remote site.

Request to Send (4): With *switched carrier operation,* an ON condition indicates to the local data set transmitter of the intent of the DTE to transmit data. With *continuous carrier,* the *carrier* is transmitted continuously. RTS may be strapped ON at the modem, or the RTS/CTS option may be exercised to provide a time delay for the transmitting data terminal.

Received Line Signal Detector (8): This circuit turns ON 45 ms after the data carrier has been received (41 ms after SQ turns ON). The 41 ms allows the adaptive equalizer at the receive site to "train" itself and get into synchronization with the received carrier. This circuit must be ON before data can be received on RD. The 1-second holdover option is to prevent loss of synchronization at the receive modem even though the reception of the carrier is interrupted for 1 second or less. During this time, any demodulated signals are passed on to the data terminal on RD even though they may be invalid. RLSD ON enables the RD pin.

Clear to Send (5): With switched carrier operation, this circuit will go ON 48.5 ms after RTS is received by the modem. This delay is built into the modem and should be greater than the time for SQ ON plus the time for the RLSD ON at the receive modem. When this circuit goes ON, the modem indicates to the DTE that it is ready to transmit data. With continuous carrier operation, if RTS is strapped ON, the modem will supply a constant ON on this line. With continuous carrier and the switched RTS option, this circuit will go ON 8 ms after the modem receives an RTS. CTS ON enables the TD pin.

Transmitted Data (2): Serial digital data in the form of marks and spaces are transmitted on this line from the DTE to the DCE. The data set samples this data stream on the ON to OFF transition of the Transmitter Signal Element Timing circuit. Data should not be transmitted on this circuit unless RTS, CTS, and DSR are ON.

Transmission Signal Element Timing (15): This circuit provides the timing for the transmission of data from the DTE to the modem. Data are sent to the modem on the OFF-to-ON transition of this line and are sampled by the modem on the ON-to-OFF transition of this line. The signal on this line is generated by an internal clock (4800 Hz) in the modem.

Transmit Signal Element Timing (24): If external (to the data set) timing is desired, the DTE can send timing information on this circuit to the modem and the internal modem clock will phase lock onto it. If the modem is designed for a 4800-bps data rate, the signal on this line should be at that frequency. The output onto TD will still be controlled by the signal from the modem on circuit 15.

Receiver Signal Element Timing (17): Clocking information on this line provides the receive DTE with timing information concerning the rate at which data are being sent from the DCE to the DTE on RD. The ON-to-OFF transition marks the center of the data bit on the RD line.

Received Data (3): Received demodulated data are sent on this line from the DCE to the DTE. If no data are being received (RLSD OFF), a mark generator internal to the modem will keep a constant mark on the RD line.

Signal Quality Detector (21): This circuit provides a fast response indication of the presence or absence of a data carrier signal from a distant terminal. This line goes ON if a carrier is received for 4 ms or longer and goes OFF if the data carrier is absent for 2 ms or longer.

Equalizer Mode (11) (non-EIA): This circuit is used by the data set to notify the DTE when the data set begins self-adjusting because error performance is poor. When RLSD is ON and this circuit is OFF, the data set is retraining and the probability of error in the data on RD is high. When RLSD is ON and this circuit is ON, the modem is trained and the probability of error on RD is small.

New Synch (14) (non-EIA): The use of this circuit is optional and is intended for use with a data set at the primary of a multipoint dedicated network. When the primary is in a polling operation, rapid resynchronization of the receiver to many remote transmitters is required. The receiver clock normally maintains the timing information of the previous message for some interval after the message has ended. This may interfere with resynchronization on the receipt of the next message. An ON condition should be applied to this circuit by the DTE for at least 1 ms but no longer than the intermessage interval to squelch the existing timing information in the modem after the message has been received.

Example   Signaling Sequence on the RS 232C Interface

The primary wishes to transmit a message to the secondary. Assume that the primary and the secondary DSR and DTR are ON. The primary modem is operating in the switched carrier mode.

The primary causes the RTS to go ON. When the modem detects this circuit ON, it starts to transmit a carrier to the secondary. After the secondary has received the carrier for 4 ms, SQ turns ON, validating the reception of the carrier, and the secondary modem starts training on this carrier. After 41 ms the secondary has completed training and RLSD goes ON. Forty-eight milliseconds after the primary modem has received the RTS it turns CTS ON to notify the primary that it can start transmitting data. At this time, the primary starts sending serial digital data to the data set on TD. These data bits phase-shift modulate the carrier and the resultant signal is transmitted. At the receive end, the modem demodulates the incoming signal converting the information back to serial digital data and sends these data to the secondary DTE on the RD line. Since the RLSD is ON, the secondary DTE accepts the information on the RD line as data. When the primary completes data transmission, it returns RTS back to OFF. When the transmit modem detects this change on RTS, it causes CTS to go OFF and stops transmitting the carrier. At the receive end, 2 ms after the carrier is lost, SQ goes OFF, followed almost immediately by RLSD going OFF. The internal mark generator of the secondary modem will now output constant marks on the RD line. This last state of the RD line is called the *mark hold condition*. While a carrier is being received by the receiving station but no data are present on the carrier, the mark generator circuit of the receive modem is disconnected from the RD line and the carrier is converted to constant marks on the RD line. As far as the receive DTE is concerned, by simply observing the constant marks on the RD line, it cannot tell how they originated. The constant marks derived from the carrier produce an *idle line 1 condition*.

Demodulation at the receive end is not instantaneous. Since the transmitter stops transmitting a carrier after the last bit of data has been sent, it is possible for the RLSD at the receive end to go OFF prior to the time that the modem has completed demodulating the last piece of received information. If the RLSD is OFF, anything placed on the RD line by the modem will not be accepted as data by the DTE. Hence the need of the 2-ms RLSD OFF delay—to allow the last piece of demodulated information to get to the DTE after the carrier is turned OFF.

There are many different types of modems. They differ in their transmission rate and in the type of modulation that they perform. The RLSD ON and OFF delays may differ considerably as well as the RTS/CTS delays.

Example Timing Diagram of the Signaling Sequence on a RS 232C Interface (Figure 6–2)

- RTS/CTS delays
  —40 ms: Switched carrier option.
  —20 ms: Continuous carrier with switched RTS option.

- RLSD
  —Turn ON 10 ms after the reception of the analog carrier. For simplicity, this time will include the SQ time.
  —Turn OFF 0 ms after the loss of the analog carrier. This time will also include the SQ time.

- The primary is operating with continuous carrier, switched RTS option.

- The secondary is operating with switched carrier.

- The primary will send a 150-ms message to the secondary.

- Propagation delay (time for a signal to go from the primary modem to the secondary modem, or vice versa) = 30 ms.

- Secondary turnaround time = 100 ms. The turnaround time is the time from which a message is received until the response is transmitted. This time can be subdivided into:
  —Terminal turnaround time: the time a terminal receives a message, checks it for errors, decides what the response should be, and composes the reply.
  —RTS/CTS delay
  In this example, these two components will be handled individually.

- The secondary responds with a 20-ms message.

- Primary and secondary data sets are turned ON at $t = 0$ ms.

- Primary turns RTS ON at $t = 20$ ms.

On the timing diagram (Figure 6–2), the transmitted carrier (TxCar) and the received carrier (RxCar) are not signals on the RS 232C interface. The carrier is transmitted on the telephone lines from modem to modem.

$t = 0$ Since the primary data set was turned ON at this time, and it is operating with continuous carrier option, it starts transmitting the carrier immediately.

$t = 30$ Since the propagation delay is 30 ms, the secondary will start receiving the carrier at this time.

$t = 40$ Ten milliseconds later, the secondary's RLSD turns ON. It took this amount of time for the secondary modem to detect the carrier and train on it.

$t = 20$ The primary terminal turns RTS ON.

$t = 40$ Twenty milliseconds later, the primary terminal receives a CTS ON from its

Figure 6-2 Timing diagram of a signal sequence on an RS 232C interface.

Primary

RTS
CTS
TD
TxCar
RxCar
RLSD
RD

Secondary

RTS
CTS
TD
TxCar
RxCar
RLSD
RD

t (ms)

Indicates the RD pin is active
although data have not reached
the receiver yet

modem. It has allowed time for the secondary modem to detect and train onto the carrier. When the primary receives CTS ON, it starts transmitting data. The HIGH on the TD line is not to be misinterpreted as an ON condition. The HIGH on this line represents the time duration during which the data are sent. During this time, the actual signals on this line will be marks and spaces.

$t = 70$ Because of the propagation delay, data are received at the secondary 30 ms after they were transmitted by the primary. Comments made previously concerning the HIGH on the TD line also apply to the RD line. When there are no data on the RD line, the signal on this line is held at a constant mark condition (mark hold) by the modem. To prevent confusion, this is not shown on the timing program.

$t = 190$ The primary completes the transmission of the 150-ms message. On completion of transmission, the primary returns RTS to OFF.

$t = 220$ Thirty milliseconds after the primary transmitted the last piece of data, the secondary receives the last piece of data.

$t = 320$ One hundred milliseconds turnaround time for the secondary. At this time, the secondary turns RTS ON and immediately starts transmitting its carrier back to the primary.

$t = 350$ The primary modem receives the secondary's carrier after the propagation delay.

$t = 360$ After 10 ms to detect and train onto the carrier, the primary modem turns RLSD ON. Forty milliseconds after the secondary modem received an RTS ON from the secondary terminal, it returns CTS ON to the terminal. The terminal, on receiving CTS ON, starts transmitting data.

$t = 390$ Thirty milliseconds after the secondary started to transmit data, the primary starts to receive the data.

$t = 380$ The secondary completes data transmission and places RTS OFF. The secondary modem, sensing this change on RTS, places an OFF on CTS and, at the same time, stops transmitting its carrier. Note that, since the primary was operating with continuous carrier option, its carrier is still being transmitted.

$t = 410$ Thirty milliseconds after the secondary stopped transmitting data, the primary stops receiving data. Thirty milliseconds after the secondary stopped transmitting the carrier, the primary stopped receiving the carrier. Since there is no RLSD OFF delay, the primary modem immediately places an OFF on RLSD. Protocols such as Bisynch terminate each transmission with a trailing pad of FFH. This trailing pad allows the receive modem sufficient time to demodulate the last piece of data and send it out on RD before RLSD goes OFF.

This example presumes a multipoint, four-wire, dedicated system. Each line in Figure 6–3 represents a two-wire pair. Since the primary is operating with continuous carrier option, all secondary station modems should be continuously trained onto it. Only one secondary station can transmit to the primary at any given time. The carrier frequency of each station may be different. Therefore, the necessity that each operate with a switched carrier option. The primary modem must be able to synchronize rapidly onto the carrier of the station that

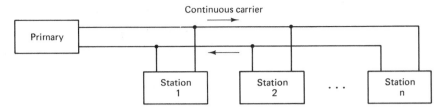

Figure 6–3 Primary: continuous carrier; secondaries: switched carrier.

is transmitting. Hence the need of a New Sync signal from the primary terminal to the modem.

RS 422A; RS 423A; RS 449

RS 232C was intended for interfacing the computer/data terminal to the modem, where the separating distance was 50 ft or less. Realizing that this standard was unsuitable where greater distances were required between the terminal and the modem, EIA has generated three new standards: RS 422A (*Electrical Characteristics of Balanced Voltage Digital Interface Circuits*), RS 423A (*Electrical Characteristics of Unbalanced Voltage Digital Interface Circuits*), and RS 449 (*General Purpose 37-Position and 9-Position Interface for Data Terminal Equipment and Data Circuit-Terminating Equipment Employing Serial Binary Data Interchange*).

From the Appendix to RS 422A (not actually a formal part of the EIA Recommended Standard):

> The maximum permissible length of cable separating the generator and load is a function of data signalling rate and is influenced by the tolerable signal distortion, the amount of longitudinally coupled noise and ground potential difference introduced between the generator and the load circuit grounds as well as by cable balance. Increasing the physical separation and interconnecting cable length between the generator and the load interface points increases exposure to common mode noise, signal distortion, and the effects of cable imbalance. Accordingly, users are advised to restrict cable length to a minimum, consistent with the generator-load physical separation requirements [see Figure 6–4].
>
> At a 10 Mbit/s signalling rate, the cable length is limited to approximately 15 m. As the data signalling rate is reduced below 90 kbit/s, the cable length has been limited at 1200 meters (4000 feet) by the assumed maximum allowable 6 dBV signal loss.
>
> For the RS 422A, terminal A of the generator shall be negative with respect to the B terminal for a binary 1 (MARK or OFF) state. The A terminal of the generator shall be positive with respect to the B terminal for a binary 0 (SPACE or ON) state.

From the Appendix to RS 423A (not a formal part of the EIA Recommended Standard): for a 3-μs linear rise time, the maximum cable length is 90 m and the maximum signaling rate 100 kbps. For approximately a 34-μs linear rise time, the maximum cable length is 600 m and the maximum signalling rate is 60 kbps. An unbalanced digital interface circuit is shown in Figure 6–5.

R$_t$ = optional cable termination resistance
V$_g$ = ground potential difference
A, B   = generator interface points
A$'$, B$'$ = load interface points
C   = generator circuit ground
C$'$ = load circuit ground

Figure 6–4   Balanced digital interface circuit. (EIA: RS 422A)

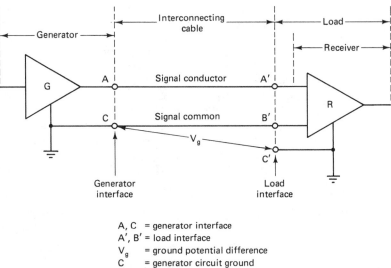

A, C   = generator interface
A$'$, B$'$ = load interface
V$_g$ = ground potential difference
C   = generator circuit ground
C$'$ = load circuit ground

Figure 6–5   Unbalanced digital interface circuit. (EIA: RS 423A)

For the RS 423A, the A terminal of the generator shall be negative with respect to the C terminal for a binary 1 (MARK or OFF) state. The A terminal of the generator shall be positive with respect to the C terminal for a binary 0 (SPACE or ON) state.

RS 449 is an updated version of the RS 232C. It has two cables, a 37-pin connector and a 9-pin connector. The 37-pin connector is used for interfacing signals, while the 9-pin connector is used for diagnostics. With the RS 449, the two-letter pin designators much more accurately define the pin functions when compared with the RS 232C.

RS 449 divides interchange circuits into two categories: category I (two data, three timing, and five other circuits) and category II (all others). Below 20 kbps, category I circuits can be implemented with RS 422A or RS 423A drivers and receivers; above 20 kbps, RS 422 circuitry must be used. Category II circuits (generally status and maintenance circuits) always use RS 423A.

RS 449 Interface*

Pin number	Mnemonic	Circuit name	Circuit category
		37-Pin Cable Primary	
1		Shield	
2	SI	Signaling Rate Indicator	II
3, 21		Spare	
4, 22	SD	Send Data	I
5, 23	ST	Send Timing	I
6, 24	RD	Receive Data	I
7, 25	RS	Request to Send	I
8, 26	RT	Receive Timing	I
9, 27	CS	Clear to Send	I
10	LL	Local Loopback	II
11, 29	DM	Data Mode	I
12, 30	TR	Terminal Ready	I
13, 31	RR	Receiver Ready	I
14	RL	Remote Loopback	II
15	IC	Incoming Call	II
16	SF/SR	Select Frequency/Signaling Rate Select	II
17, 35	TT	Terminal Timing	I
18	TM	Test Mode	II
19	SG	Signal Ground	II
20	RC	Receive Common	
28	IS	Terminal in Service	II
32	SS	Select Standby	II
33	SQ	Signal Quality	II

Pin number	Mnemonic	Circuit name	Circuit category
34	NS	New Signal	II
36	SB	Standby Indicator	II
37	SC	Send Common	II
		9-Pin Cable Secondary	
1		Shield	
2	SRR	Secondary Receiver Ready	II
3	SSD	Secondary Send Data	II
4	SRD	Secondary Receive Data	II
5	SG	Signal Ground	II
6	RC	Receive Common	II
7	SRS	Secondary Request to Send	II
8	SCS	Secondary Clear to Send	II
9	SC	Send Common	II

*(Courtesy Electronic Industries Association, EIA. This document, RS 449, is available in its entirety from the EIA, Washington, D.C.)

Much of the field equipment is presently RS 232C compatible. As the RS 449 replaces the RS 232C, problems will arise when the modem connectors are made for the RS 449 while the terminal equipment satisfies RS 232C requirements. Adapters are available through which an RS 232C terminal can hook onto an RS 449 interface.

As EIA sets standards in the United States, CCITT (Comité Consultatif International Téléphonique et Télégraphique) sets standards in the European countries. V.28 (*Electrical Characteristics for Unbalanced Double-Current Interchange Circuits*) by CCITT parallels the voltage specifications of RS 232C, and V.24 (*List of Definitions for Interchange Circuits between Data Terminal Equipment and Data Circuit Terminating Equipment*) parallels the pin designations of the RS 232C. X.20 concerns itself with the *Interface between Data Terminal Equipment for Start-Stop Transmission Services on Public Data Networks*. European standards are readily recognizable in American literature since the more common standards are identified as either X.xx or V.xx, where xx is a two-digit number.

20-mA LOOP

The 20-ma loop is often called the teletype (TTY) interface. Originally, it was used to interface mechanical teleprinters to the telephone line. A teleprinter is a combination of printer and keyboard. Video displays (CRTs) are not included. Interface arrangements that transfer signals with current flow rather than voltage levels (such as the 20-ma loop) are generally used to control switch closures in mechanical devices. With the 20-mA loop, a logic 1 or mark condition is indicated by the presence of the dc current, and a logic 0 or space is indicated by the absence of this current. Continuity is monitored during idle line conditions by simply testing for current on the line. This is believed to be why data communication circuits transmit idle line 1's in the absence of data.

Today, the 20-mA loop is used occasionally to interface microprocessors and other digital equipment to external line printers. Due to the mechanical nature of the printers and teleprinters, they generate noise spikes that could interfere with the data or possibly damage delicate equipment that may be attached to the other end of the interface. Very often, optical couplers are used to isolate the equipment and suppress the noise spikes.

Teleprinters are rapidly becoming extinct; they are being replaced by modern keyboard displays and high-speed printers. Consequently, the 20-mA loop is seldom used today and the RS 232C has become the standard interface.

REFERENCES

EIA Standard RS 232C, *Interface between Data Terminal Equipment and Data Communication Equipment Employing Serial Binary Data Interchange,* August 1969.

EIA Standard RS 422A, *Electrical Characteristics of Balanced Voltage Digital Interface Circuits,* December 1978.

EIA Standard RS 423A, *Electrical Characteristics of Unbalanced Voltage Digital Interface Circuits,* December 1978.

EIA Standard RS 449, *General Purpose 37-Position and 9-Position Interface for Data Terminal Equipment and Data Circuit-Terminating Equipment Employing Serial Binary Data Interchange,* November 1977.

The references above may be obtained from:

Electronic Industries Association
2001 Eye Street, N.W.
Washington, DC 20006

QUESTIONS

1. What is the nominal maximum physical length of the RS 232C interface cable?
2. What characteristics of the RS 232C interface cable limit its physical length?
3. The Clear to Send (CTS) signal is a handshake from the DCE to the DTE in response to the _____ signal.
4. What is the purpose of the RTS/CTS delay?
5. What signal on the RS 232C interface must be active before the RTS can go active?
6. What signal on the RS 232C interface indicates the presence of a received carrier?
7. The receive data (RD) pin of the RS 232C interface is enabled by an active condition of the _____ signal.
8. + 10 V dc on a data pin represents a logic _____ .

9. -12 V dc on a control pin represents a(an) _____ condition.

10. A driver transfers signals from the DTE to the DCE and never from the DCE to the DTE. (T, F)

11. What is the maximum terminated voltage that should be present on the RS 232C interface?

12. Analog signals are present (*between the DTE and the DCE, between the LCU and the DTE, between the transmit DCE and the receive DCE*).

13. The RS 232C interface specifies a _____ pin cable.

14. The RS 232C interface is between the (*LCU and the DTE, DTE and the DCE, LCU and the UART, transmit and receive modems*).

15. With switched carrier operation, the transmit modem will begin transmitting an analog carrier when _____ goes active.

16. RTS/CTS will be optioned (*on, off*) for switched carrier operation.

17. The RS 232C is gradually being replaced by the _____ .

18. When transmitting a message, first the (*CTS goes active, RTS goes active, DSR must be active, RLSD goes active*).

19. The RS 449 interface is comprised of a _____-wire cable for interface signaling and a _____-wire cable for diagnostics.

20. The RS 422A and RS 423A standards specify the physical limitations for the RS 449 standard. (T, F)

21. The RS 422A standard is a (*balanced, unbalanced*) cable.

22. The maximum possible physical length of the RS 449 interface is (*less than, equal to, greater than*) that of the RS 232C interface.

23. The maximum kbps rate for the RS 423A standard remains constant regardless of the physical length. (T, F)

24. What is the European counterpart of the EIA?

PROBLEMS

1. For the following modem and circuit specifications, complete the timing diagram in Figure P6–1 for the RS 232C signaling sequence.

   - RTS/CTS delay:
     —40 ms: switched carrier option
     —20 ms: continuous carrier option with switched RTS
   - RLSD delay:
     —20 ms: turn on
     —10 ms: turn off
   - Primary optioned for switched carrier, switched RTS
   - Secondary optioned for switched carrier, switched RTS
   - Propagation time: 20 ms in both directions
   - Secondary turn around time: 40 ms
   - Primary message length: 80 ms
   - Secondary message length: 40 ms
   - Primary begins sequence at time $t = 0$ ms (RTS goes active)

Primary
RTS
CTS
TD
TxCar

RxCar
RLSD
RD

Secondary
RTS
CTS
TD
TxCar

RxCar
RLSD
RD

0 20 40 60 80 100 120 140 160 180 200 220 240 260 280 300 320 340 360 380 400 420 440 460 480 500

t (ms)

Figure P6–1

Primary
RTS
CTS
TD
TxCar

RxCar
RLSD
RD

Secondary
RTS
CTS
TD
TxCar

RxCar
RLSD
RD

0 20 40 60 80 100 120 140 160 180 200 220 240 260 280 300 320 340 360 380 400 420 440 460 480 500

Figure P6–2

2. For the following modem and circuit specifications, complete the timing diagram in Figure P6–2 for the RS 232C signaling sequence.

- RTS/CTS delay:
  —20 ms: switched carrier option
  —10 ms: continuous carrier option with switched RTS
- RLSD delay:
  —10 ms: turn on
  — 0 ms: turn off
- Primary optioned for continuous carrier, switched RTS
- Secondary optioned for switched carrier, switched RTS
- Propagation time: 40 ms in both directions
- Secondary turnaround time: 20 ms
- Primary message length: 60 ms
- Secondary message length: 20 ms
- Primary and secondary data sets both turned on at time $t = 0$ ms
- Primary begins sequence at time $t = 20$ ms (RTS goes active)

7

DATA TRANSMISSION
WITH ANALOG CARRIERS

Data transmission involves the transmittal of digital information from one DTE to another. Rather than construct special lines for this purpose, it is more convenient to use the public telephone network (PTN). Unfortunately, the vast majority of the PTN was installed long before the advent of large-scale data communications and was designed for conveying information that is analog in nature. In addition, the analog information is restricted to the frequency range from 300 to 3000 Hz. The primary function of the data set (modem) at the transmit end is to convert digital pulses to analog signals suitable for transmission on the PTN and, at the receive end, to convert the analog signals back to digital information.

Modems are categorized as being either synchronous or asynchronous. If a communications data link uses synchronous modems, the internal clocks of the transmit and the receive modems must be synchronized. These clocks are used to control the rate at which digital data flows. The type of modulation used with synchronous modems is such that the receive modem can extract clocking information from the received signal and use it to maintain synchronization with the transmit modem. In addition, carrier frequency synchronization between the transmit and receive modems must be established and maintained in order to demodulate the signal correctly at the receive end. Asynchronous modems do not have to abide by these restrictions. Clocking information is never transmitted between asynchronous modems.

Three properties of an analog signal may be varied: amplitude, frequency, and phase. Varying only the amplitude of the analog signal at a digital rate is called *amplitude shift keying* (ASK). ASK has a number of inherent disadvantages, with poor noise immunity being the most predominant; consequently, it is seldom used in practice. *Frequency shift keying* (FSK) varies the frequency at a digital rate and is used for low-speed data transmission. Modems using FSK are generally asynchronous and can operate at a maximum transmission rate of 1800 bps. Medium-speed modems are generally synchronous, operate at speeds up to 4800 bps, and use *phase shift keying* (PSK). High-speed modems (9600 bps) use *quadrature amplitude modulation* (QAM), in which both the amplitude and phase of the analog carrier are varied by the digital modulating signal. These facts can be ascertained from Table 7–1, which shows the characteristics of commonly used Bell modems.

FREQUENCY SHIFT KEYING

FSK is the process of shifting (modulating) an analog carrier frequency at a digital rate. FSK closely resembles standard frequency modulation. With FSK the modulating signal is not a sinusoidal signal but a series of dc pulses that vary between two discrete voltage levels. The output of an FSK modulator can be considered a step function in the frequency domain.

202T Modem

The Bell System model 202T modem is an asynchronous transceiver utilizing FSK modulation. It employs a 1700-Hz carrier that can be shifted at a maximum rate of 1200 times a second. When a logic 1 (mark) is applied to the modulator, the carrier is shifted down 500 Hz to 1200 Hz. When a logic 0 (space) is applied, the carrier is shifted up 500 Hz to 2200 Hz. Consequently, as the digital signal alternates between 1 and 0, the carrier is shifted back and forth between 1200 and 2200 Hz. This process is analogous to a conventional FM system. The difference between the mark and space frequencies (1200 to 2200 Hz) relates to the peak-to-peak frequency deviation, and the digital rate of change relates to the frequency of the modulating input signal. Therefore, we can say the 1700-Hz carrier is frequency-modulated by a square-wave input signal. A figure of merit used to express the degree of modulation achieved in an FSK modulator is the h factor, which is defined as:

$$h = \frac{\mid f_m - f_s \mid}{\text{bps}}$$

where f_m = mark frequency
f_s = space frequency
bps = input bit rate (bps)

For the 202T modem,

$$h = \frac{\mid 1200 - 2200 \mid}{1200} = \frac{1000}{1200} = 0.83$$

As a general rule and for best performance, the h factor is limited to a value less than 1. Note the similarity of the expression for h and the expression for the modulation index (MI) in conventional FM.

TABLE 7–1 COMMONLY USED BELL MODEMS

Bell designation	Line facility	Operating mode	Synchronization	Type of modulation	Maximum data rate (bps)	Function
103A2, A3, E	Dial-up	FDX	Asyn.	FSK	300	Originate/answer
103F	Leased (2-wire)	FDX	Asyn.	FSK	300	Originate/answer
113A	Dial-up	FDX	Asyn.	FSK	300	Originate only
113B	Dial-up	FDX	Asyn.	FSK	300	Answer only
201B	Dial-up	HDX/FDX	Syn.	4PSK	2400	
201C	Leased (4-wire)	HDX/FDX	Syn.	4PSK	2400	
202S	Dial-up	HDX	Asyn.	FSK	1200	
202T	Leased (4-wire)	HDX/FDX	Asyn.	FSK	1200 (Basic, 3002 channel) 1800 (C1 conditioning)	
208A	Leased (4-wire)	HDX/FDX	Syn.	8PSK	4800	
208B	Dial-up	HDX	Syn.	8PSK	4800	
209A	Leased (4-wire)	FDX	Syn.	QAM	9600 (D1 conditioning)	

$$MI = \frac{\Delta f}{f_{mod}}$$

where Δf = peak frequency deviation
         f_{mod} = modulating signal frequency

The modulation index for the 202T modem is

$$\Delta f = \frac{|f_m - f_s|}{2} = \frac{|1200 - 2200|}{2} = 500 \text{ Hz}$$

This converts the peak-to-peak frequency swing of the modem to a peak frequency deviation:

$$f_{mod} = \frac{f_b}{2} = \frac{1200}{2} = 600 \text{ Hz}$$

where f_b is the input bit rate (bps).

$$MI = \frac{500}{600} = 0.83$$

which is the same as the h factor.

Why does $f_{mod} = \frac{f_b}{2}$? The fastest transitions by a 1200-bps digital signal would produce a square wave of 600 Hz. The time of each alteration would be 1/1200 s (Figure 7–1).

Fourier analysis demonstrates that any periodic wave is made up of the sum of a series of harmonically related sine waves. A square wave is a periodic wave made up of the fundamental frequency and an infinite number of odd harmonics. The fundamental frequency is equal to $1/2 t_b$ (t_b = time of 1 bit).

With FM the number of side frequencies generated is directly related to the modulation index. The separation between adjacent side frequencies is equal to the frequency of the modulating signal. In the case of the 202T modem, the frequency spectrum of Figure 7–2 would result. Any additional side frequencies beyond the second set would extend beyond the passband of a basic voice band telephone channel and would be lost in transmission. Consequently, it is desirable to concentrate as much

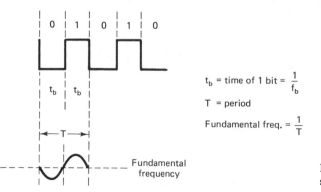

t_b = time of 1 bit = $\frac{1}{f_b}$

T = period

Fundamental freq. = $\frac{1}{T}$

Figure 7–1 Fundamental frequency of a square wave.

Figure 7–2   Frequency spectrum of a 202T modem.

energy as possible within the 300- to 3000-Hz passband. An abbreviated Bessel function chart (Table 7–2) illustrates the significant side frequencies and their relative amplitudes for a given modulation index. Note that if the modulation index is less than 1, no significant side frequencies beyond the second set are produced. Therefore, no energy is wasted in frequency components beyond the passband of the telephone channel.

In the previous analysis, the output frequency spectrum of a 202T modem was derived for an input bit rate of 1200 bps. What if the input bit rate were less than 1200 bps?

Example

Input bit rate = 500 bps.

$$\text{MI} = \frac{\Delta f}{f_{\text{mod}}} = \frac{500}{250} = 2$$

The peak frequency deviation (Δf) is still 500 Hz, because this is a function of the amplitude of the input signal and this has remained unchanged. f_{mod} is a sine wave equal to one-half of the bit rate, as explained previously. From the Bessel function chart, a modulation index of 2.0 creates four pairs of significant side frequencies. These side frequencies are separated by 250 Hz (the modulating signal frequency). This frequency spectrum is shown in Figure 7–3.

From this we can conclude that *as the modulating frequency decreases, the bandwidth decreases.* There are more frequencies present in the modulated signal, but they are spaced closer together. The worst case (widest bandwidth) occurs with the highest modulating frequency ($f_{\text{mod}} = 600$ Hz, $f_b = 1200$ bps). This worst-case modulation index is called the deviation ratio and is numerically equal to

$$\text{deviation ratio} = \frac{\Delta f \text{ (max)}}{f_{\text{mod}} \text{ (max)}}$$

TABLE 7–2   BESSEL FUNCTION CHART

MI	J_0	J_1	J_2	J_3	J_4
0.0	1.00				
0.25	0.98	0.12			
0.5	0.94	0.24	0.03		
1.0	0.77	0.44	0.11	0.02	
1.5	0.51	0.56	0.23	0.06	0.01
2.0	0.22	0.58	0.35	0.13	0.03

Figure 7–3   Frequency spectrum for a 202T modem with an input bit rate of 500 bps.

What about the sine-wave components present in the input square wave that are the odd harmonics of the fundamental frequency? Fourier analysis of a square wave indicates that the amplitudes of the nth harmonics decrease rapidly as n increases. Consequently, any frequency deviation caused by them is insignificant and need not be considered. Most of the side frequencies generated by these high harmonics produce frequency components that extend beyond the passband of the telephone channel and are lost anyway.

The 202T modem is a four-wire, asynchronous modem designed to be used with four-wire, dedicated, private-line circuits. Because the transmit and receive analog signal paths are physically separated and isolated from each other, full-duplex operation is possible with no interference between the two signals. A block diagram of the 202T transmitter is shown in Figure 7–4.

The transmitting DTE initiates the message transfer by applying an ON signal to the Request to Send (pin 4) of the RS 232C interface. At this time, the VCO begins transmitting the mark frequency (1200 Hz). After a predetermined time delay, the modem responds by applying an ON signal to Clear to Send (pin 5) of the interface. At this time, Transmit Data (pin 2) is enabled, and data from the DTE are allowed to FSK modulate the VCO. The amplifier and the bandpass filter provide sufficient amplification and bandlimiting to ensure that maximum allowable signal power (0 dBm) is applied to the Telco line within the 300- to 3000-Hz bandwidth.

Receiver designs vary with the manufacturer. There are a number of acceptable methods currently being used to produce satisfactory FSK demodulation. These techniques include the use of phase-locked loops (PLL), frequency-to-voltage converters, and tone detectors.

In the receive section (Figure 7–5), the bandpass filter limits the input FSK

Figure 7–4   Low index FSK transmitter.

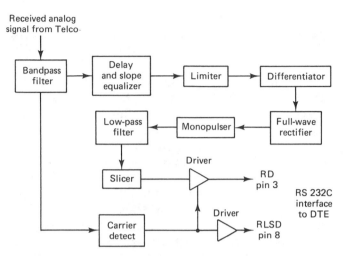

Figure 7–5   202T receiver block diagram (FSK) (limited half-cycle detection).

modulated signal to the 300- to 3000-Hz passband. The slope and delay equalizers provide post-equalization to compensate for long-term variations in the Telco transmission-line parameters. The equalizers may be manually or automatically adjusted. The limiter is an overdriven amplifier that provides an output square wave at either the mark or space frequency. The differentiator, full-wave rectifier, and monopulser (one-shot multivibrator) convert the square wave into a rectangular wave at twice the mark and space frequencies. The low-pass filter converts the rectangular pulses into a dc voltage and applies this voltage to the slicer, where the logic condition (1, 0) is determined. The carrier detect circuit indicates to the receive DTE when the modem is receiving an analog carrier on RLSD (pin 8) of the interface. This circuit also enables the RD driver (pin 3) and allows received data to be transferred to the DTE.

202S Modem

All data communication systems do not require the conveniences offered by dedicated lines. The data through-put requirements of a particular user may not justify the expense of a private line, or the type of data being transferred may not require the reliability. An affordable alternative to private-line subscription is the use of the already established direct distance dialing (DDD) network. An advantage of a dial-up system is that the subscribers pay only for the time they are actually using the Telco facilities. The user can initiate a call when there is a message to send, terminate the call upon completion, and pay only for the time of the call. The disadvantage is that the user must contend for a circuit with the rest of the network (and there are millions of other users). Also, each telephone call established could be made over different Telco facilities that do not necessarily have the same performance parameters. Consequently, the reliability (error performance) of a dial-up connection is not consistent from call to call.

A major restriction that must be considered when designing modems for dial-up applications is that they must be capable of functioning over two-wire lines. The DDD network was designed for voice communications that requires only two wires for satisfactory performance. If duplex operation is necessary or desired, both transmission and reception must be accomplished on a single pair of wires.

The Bell System 202S modem was designed for two-wire DDD applications. The mark/space frequency assignment, information rate (bps), and modulation scheme (FSK) are identical to that of the 202T (four-wire version). A mark is represented by a 1200-Hz tone and a space by a 2200-Hz tone (see Figure 7–6). The maximum bit rate is 1200 bps. Half-duplex operation is accomplished by disabling the receiver while transmitting and disabling the transmitter while receiving. Full-duplex operation is possible only through the use of a secondary reverse or backward channel. For the secondary channel not to interfere with the primary channel, a different carrier frequency must be used. The secondary amplitude shift keys a 387-Hz carrier at a very low rate (5 bps). *ASK* is simply turning the carrier on and off at a digital rate. AM sidebands are generated due to the AM nature of the modulation process, but because of the extremely low bit rate, the bandwidth required for transmission is very narrow. This process of sharing a limited bandwidth on a single facility is called *multiplexing*. Because the frequency domain of the existing channel is shared, the process is called *frequency-division multiplexing* (FDM). The 300- to 3000-Hz bandwidth of the telephone channel is divided between the primary and the secondary channels. The primary channel uses most of the bandwidth for message transfer, while the secondary channel occupies only a narrow band at the low end of the spectrum for transmission of data link control signals.

Figure 7–6 Output frequency spectrum of a 202S modem.

| 387 | 1200 | 2200 Hz |
| | F_M | F_S |

103 Modem

A modem more popular than the 202S for dial-up applications is the Bell System 103 modem. This is a 300-bps, asynchronous modem, designed for full-duplex operation. It also uses FDM to achieve full duplex. In this case, the 300- to 3000-Hz bandwidth of the telephone channel is divided in half; the upper band extends from 1650 to 3000 Hz and the lower band extends from 300 to 1650 Hz. Because each channel is bandlimited to one-half of the original bandwidth, the information rates possible with the 103 modem are less than that of the 202S. Each channel has its own mark and space frequencies. The mark/space frequencies associated with the lower channel are called the *originate frequencies,* while those associated with the upper channel are called the *answer frequencies.* The modem that initiates the call transmits the originate frequencies and receives the answer frequencies. The modem that terminates or answers the call transmits the answer frequencies and receives the originate frequencies.

When the modem is optioned or identified to operate in the originate or answer mode, the transmit frequencies of that modem are also identified. The originate and answer frequencies are referred to as the *F1* and *F2 frequencies,* respectively (Figure 7–7).

| 1070 | 1270 | 1650 | 2025 | 2225 Hz |
| F1$_S$ | F1$_M$ | | F2$_S$ | F2$_M$ |

Originate   Answer

Figure 7–7   Output frequency spectrum of the 103 modem.

Spectral analysis of the 103 modem:

$$h = \frac{|F1_m - F1_s|}{bps} = \frac{|1270 - 1070|}{300} = 0.67$$

$$h = \frac{|F2_m - F2_s|}{bps} = \frac{|2225 - 2025|}{300} = 0.67$$

$$\text{deviation ratio} = \frac{\Delta f\,(max)}{f_{mod}\,(max)} = \frac{100}{150} = 0.67$$

$$f_{mod} = \frac{f_b(bps)}{2}$$

A modulation index of 0.67 yields two sets of significant side frequencies centered around the low channel carrier of 1170 Hz and the high channel carrier of 2125 Hz (see Figure 7–8). On a two-wire facility, if full-duplex operation was used at a bit rate of 300 bps or less, no interference between the two channels would be encountered.

| 870 | 1020 | 1170 | 1320 | 1470 | 1650 | 1825 | 1975 | 2125 | 2275 | 2425 Hz |

Low channel   High channel

Figure 7–8   Frequency spectrum of the high and low channels of a 103 modem with an input bit rate of 300bps.

PHASE SHIFT KEYING

Binary Phase Shift Keying

Phase shift keying (PSK) is a form of modulation by a modem where the phase of the carrier is changed by the binary data. Although the data input to the modem conforms to RS 232C electrical standards, the internal circuitry of the modem can convert these voltage values to levels necessary to perform the modulation. In BPSK (binary phase shift keying), the modulator output has only two phases: it is either in phase or 180° out of phase, with the carrier oscillator depending on whether the input is a logic 1 or 0, respectively. Since a balanced modulator produces a product of its input signals, if a logic 1 is assigned a positive voltage and a logic 0 a negative voltage, the output of the modulator circuit in Figure 7–9 is either cos $\omega_c t$ or $-$cos $\omega_c t$. In the receiver demodulator circuit of Figure 7–10, the carrier is detected from the incoming signal and then mixed with the incoming signal to produce either

$$\cos^2 \omega_c t = \frac{1}{2}(1 + \cos 2\omega_c t)$$

or

$$-\cos^2 \omega_c t = -\frac{1}{2}(1 + \cos 2\omega_c t)$$

The low-pass filter at the output eliminates the double-frequency component, leaving only a dc value which is of the same polarity as the original modulating signal (data bit). The signal can be amplified to the level necessary to be coupled to the DTE by way of the RS 232C interface.

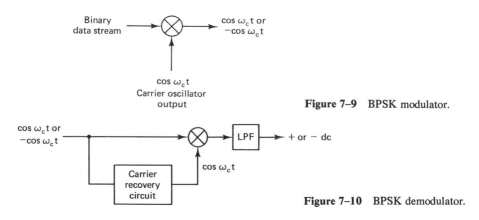

Figure 7–9 BPSK modulator.

Figure 7–10 BPSK demodulator.

Ring modulator.   One method of achieving BPSK modulation is with a balanced modulator. The ring modulator in Figure 7–11 is one form of balanced modulator. The major characteristic of a balanced modulator is that neither of the input frequencies are present at the output—only their product. For the ring modulator to function properly, the voltage applied to the X-Y input must be greater than the voltage applied at the R-S input. The voltage applied to the X-Y input always determines which diodes are conducting. The voltage at the R-S input cannot turn off a diode turned on by the X-Y voltage, nor can the R-S voltage turn on a diode held off by the X-Y voltage. If only the X-Y voltage is applied with the polarity shown in Figure 7–12, only diodes A and D will be conducting. Since this will produce equal and opposite currents in each half of the primary of the output transformer, no voltage will be coupled to the output. If the X-Y voltage is reversed as shown in Figure 7–13, diodes B and C will now conduct, but again, there will be no output voltage. If only a voltage is applied to the R-S input, either diodes A and B or diodes C and D will effectively short the input from the output.

Figure 7–11   Ring modulator.

Figure 7–12   X-Y input only to the ring modulator.

Figure 7–13   X-Y input only to the ring modulator.

This is shown in Figure 7–14. This circuit is used to produce double-sideband suppressed carrier (DSBSC) in AM with the carrier applied to the X-Y input and the modulating signal applied to the R-S input.

For this circuit to be used as a BPSK modulator, the oscillator output is applied to the R-S input and the input data stream is applied to the X-Y input. With the data polarity as shown in Figure 7–15, diodes A and D will be on. During the positive polarity of the oscillator voltage, diode A will conduct more than diode D. This produces a greater current in the upper half of the output primary coupling a positive

Figure 7-14 R-S input only to the ring modulator.

Figure 7-15 In-phase output signal—BPSK.

voltage to the secondary. During the negative alternation of the oscillator voltage, diode D will conduct more heavily coupling a negative voltage to the output. With the data voltage as shown in Figure 7–15, the output frequency is in phase with the oscillator frequency.

If the data polarity is reversed as shown in Figure 7–16, diodes B and C will conduct. During the positive alternation of the oscillator voltage, diode B conducts more heavily, producing a greater current in the upper half of the output primary and couples a negative voltage to the output. During the negative alternation of the oscillator voltage, diode C conducts more heavily producing a greater current through

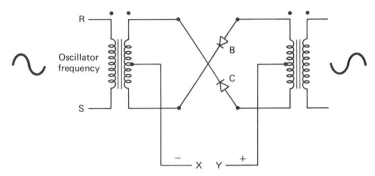

Figure 7–16   Out-of-phase output signal—BPSK.

the lower half of the output primary and couples a positive voltage to the output. With the data voltage as shown in Figure 7–16, the output voltage will be 180° out of phase with the oscillator voltage. Therefore, by varying the polarity of the voltage at the X-Y input, two phases, 180° apart, are generated as required by the BPSK modulator.

Quadrature PSK

With quadrature PSK (QPSK or 4-PSK) the output signal may have one of four possible phases. The output phase is determined by considering the status of two bits, called *dibits*. A block diagram of a QPSK modulator is shown in Figure 7–17. The four phases are 90° apart. The bit splitter alternately divides the input bit stream into two channels, the I (*in-phase*) channel and the Q (*quadrature*) channel. The bits entering the I channel are mixed with a carrier frequency that is in phase with the reference carrier oscillator, and the bits entering the Q channel are mixed with a carrier frequency that is 90° out of phase (in quadrature) with the reference carrier oscillator. The bit rate of the I and Q channels is equal to one-half of the input bit stream. Since bits *a* and *b* can

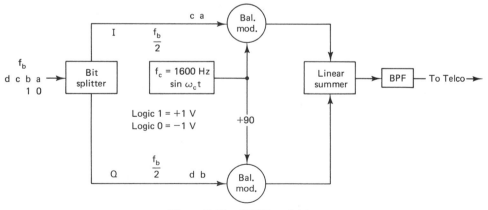

Figure 7–17   QPSK modulator.

be either a logic 1 or 0, the output of the top balanced modulator can be either $+\sin \omega_c t$ or $-\sin \omega_c t$ and the output of the bottom balanced modulator can be either $+\cos \omega_c t$ or $-\cos \omega_c t$. The four possible output phases are shown graphically in Figure 7–18.

For the input bits a and b shown in Figure 7–17, the output is $+\cos \omega_c t - \sin \omega_c t$. This output will be carried through the demodulator to recover the two original bits. Figure 7–19 shows a block diagram of a QPSK demodulator. The carrier recovery circuit recovers the original transmit carrier frequency. This carrier is mixed directly with the received signal in the top channel. The output of the balanced modulator is

$$(\sin \omega_c t)(\cos \omega_c t) - \sin^2 \omega_c t$$

$$= \frac{1}{2} \sin 2\omega_c t + \frac{1}{2} \sin(\omega_c t - \omega_c t) - \frac{1}{2}(1 - \cos 2\omega_c t)$$

The double-frequency components are filtered out by the LPF. The remaining negative voltage, which represents a logic 0, is sent to the bit combiner. The output of the bottom balanced modulator is

$$\cos^2 \omega_c t - (\sin \omega_c t)(\cos \omega_c t)$$

$$= \frac{1}{2}(1 + \cos 2\omega_c t) - \frac{1}{2} \sin 2\omega_c t - \frac{1}{2} \sin(\omega_c t - \omega_c t)$$

Again, only a positive voltage remains at the output of the LPF after the double-frequency component is removed. The recovered positive voltage represents a logic 1. The bit combiner will sequence the input bits in their proper order and output the recovered 1/0 onto the RD line.

A common transmission rate for QPSK is 2400 bps (f_b). Therefore, in the QPSK modulator, the I and Q channels have a bit rate of 1200 bps. The fastest change in this bit stream occurs with an alternating 1/0 pattern. This would represent a square wave with a frequency of 600 Hz. A square wave is made up of a fundamental frequency and

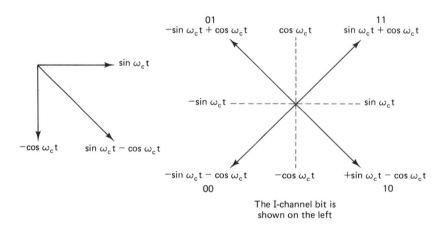

Figure 7–18 Output phasors for QPSK.

Figure 7–19   QPSK demodulator.

One bit
1200 bps

One cycle (600 Hz)

Fundamental frequency

Sampling times

Figure 7–20   Fundamental frequency of a square wave.

an infinite number of odd harmonics. The balanced modulator must modulate the carrier with all of these frequencies, producing an upper and lower side frequency for each modulating component. If only the fundamental frequency is recovered, by sampling at the correct time, the two original bits can be recovered (Figure 7–20). The side frequencies produced by the fundamental frequency in the modulation process are 2200 Hz ($f_c + f_m$) and 1000 Hz ($f_c - f_m$). The minimum bandwidth required to pass these frequencies is 1200 Hz (2200 − 1000). The purpose of the bandpass filter at the output of the linear summer is to exclude all other side frequencies from transmission.

Bandwidth efficiency.   In general, the higher the data transmission rate, the greater the bandwidth required. Data transmission rates of 2400 bps and 4800 bps have fundamental frequencies of 1200 Hz and 2400 Hz, respectively. If these frequencies amplitude-modulate the same carrier in a balanced modulator, the 1200-Hz signal would require a 2400-Hz bandwidth and the 2400-Hz signal would require a 4800-Hz bandwidth (Figure 7–21). This is called the *double-sided Nyquist* or IF bandwidth.

If the modem output is transmitted over a basic 3002 channel, the frequencies transmitted must be restricted to those allowed for voice (300 to 3000 Hz). Unless more complex modulation schemes are used, bit rates above 2400 bps quickly exceed these limits. *Bandwidth* (BW) *efficiency* is a measure of how effectively a given bandwidth is used for data transmission.

Figure 7–21   BW requirements for AM signals.

$$\text{BW efficiency (bps/Hz)} = \frac{\text{data transmission rate (bps)}}{\text{BW required (Hz)}}$$

For BPSK, since the bandwidth required is equal to the transmission rate, its BW efficiency is 1 bps/Hz. With QPSK, only a 1200-Hz BW is required for a transmission rate of 2400 bps.

$$\text{BW efficiency} = \frac{2400 \text{ bps}}{1200 \text{ Hz}} = 2 \text{ bps/Hz}$$

Because of the modulation scheme used with QPSK, the data transmission rate can be doubled with no increase in bandwidth requirements as compared to BPSK.

Baud.   Baud is a term originally used with telegraphy. It is the reciprocal of the time of the fastest signaling element that is applied *to the telephone line*. Since baud is a reciprocal of time, it is a rate. With BPSK, the times of all signaling elements are the same and are equal to the time that a constant phase is outputted by the modem as a result of modulation by a single bit. Therefore, with BPSK, the output baud is equal to the input data transmission rate. Transmission of 1200 bps to a BPSK modem is equivalent to the modem outputting a 1200-baud signal onto the telephone line. With QPSK, the time of a signaling element is the time that one of four possible phases is outputted by the modem. Since 2 bits are required to determine this phase, the output baud is one-half of the input bit rate (Figure 7–22).

The output baud may be equal to or less than the input bit rate, but it can never exceed the bit rate. The latter would infer that the signaling elements, which are determined by the incoming bits, can be generated faster than the bits are received.

Nyquist filter.   Nyquist theorems on minimum bandwidth transmission show that f_s independent symbols can be passed through a low-pass filter which has a cutoff frequency of $f_s/2$. For a low-pass filter, the cutoff frequency is also equal to the bandwidth of the filter. These Nyquist filters are ideal and are referred to as *brick-wall filters* because of their sharp cutoff characteristics (Figure 7–23). For a bandpass filter, $f_N = f_s$ (symbol rate).

Two bits at 4800 bps
T = time of 1 bit

Constant phase
t = time of one signaling
element

Figure 7–22 QPSK: bit time versus
signal element time.

f_N (Nyquist frequency)

Figure 7–23 Nyquist low-pass
brick-wall filter.

Example

QPSK—4800-bps data rate. A symbol could be represented by one of four possible phases. Since 2 bits are required to determine one symbol:

$$f_s = \frac{4800 \text{ bps}}{2} = 2400 \text{ symbols/second}$$

$$f_N \text{ for BPF} = f_s = 2400 \text{ Hz}$$

8-PSK

With 8-PSK, the modem output signal may be one of eight possible phases. The output phase is determined by considering the status of 3 bits, called *tribits*. A block diagram of an 8-PSK modulator is shown in Figure 7–24. The bit splitter divides the incoming bit stream from the DTE into three channels. Bits *a* and *b* determine the polarity of the voltage at the output of the 2-to-4 level converters while bit *c* determines the dc magnitude. Two magnitudes are used, 0.34 V and 0.821 V. The reason for these values will become obvious later. When *a* and *b* are logic 1's, the outputs of the 2-to-4 level converters are positive voltages; when *a* and *b* are logic 0's, the converter outputs are

Figure 7–24 8-PSK modulator.

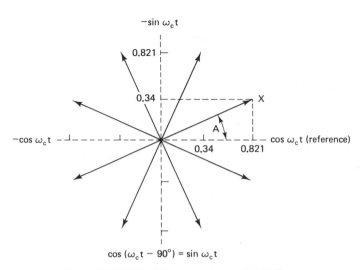

Figure 7–25   Possible output phases of 8-PSK.

negative voltages. The magnitude of the output voltages from the two converters will always be different. Whichever converter receives a logic 1 from c or $\bar{c}$ will have an output magnitude of 0.821 V; the other converter will have an output magnitude of 0.34 V. Since c and $\bar{c}$ are complements, they both cannot have the same magnitude at the same time. Since the 3 bits a, b, and c are independent of one another, ± 0.821 V and ± 0.34 V are the four possible magnitudes at the output of both converters. The output of the I-channel converter modulates the carrier directly while the output of the Q channel modulates a quadrature component of the oscillator output. The possible outputs of the I-channel balanced modulator are $\pm 0.34 \cos \omega_c t$ and $\pm 0.821 \cos \omega_c t$. The possible outputs of the Q-channel balanced modulator are $\pm 0.34 \sin \omega_c t$ and $\pm 0.821 \sin \omega_c t$. Again, the only restriction is that the two channels will always have different magnitudes. The linear summer combines the I- and Q-channel signals. The various possible phases that may be produced are shown in Figure 7–25.

$$\text{angle } A = \tan^{-1} \frac{0.34}{0.821} = 22.5°$$

This makes all phasors equally spaced and 45° apart. The numbers 0.34 and 0.821 are not sacred and could be any values as long as their ratio equaled tan 22.5°. Their amplitudes will, of course, determine signal strength.

Example

For the input bit combinations shown in Figure 7–24:

I-channel converter output: $+0.821$
Q-channel converter output: -0.34
I-channel modulator output: $+0.821 \cos \omega_c t$
Q-channel modulator output: $-0.34 \sin \omega_c t$
Output of linear summer: $0.821 \cos \omega_c t - 0.34 \sin \omega_c t$

The phasor produced by this sequence of bits is identified by X in Figure 7–25.

Depending on the input bit stream to the 2- to 4-level converter, the output levels could be either of the waveforms shown in Figure 7–26. T_{3b} is the time of three bits. The levels change at one-third the input bit rate. If each level is considered a symbol, the symbol rate at the converter output is 1600 symbols/second. To ensure information recovery, the highest fundamental frequency contained in the converter output must be represented in the transmitted signal. The highest fundamental frequency is contained in the waveform of Figure 7–26(b) and is equal to one-half of the symbol rate or 800 Hz. If all of these frequencies amplitude-modulated the carrier, the bandwidth required for transmission would range from $f_c - 800$ Hz (900 Hz) to $f_c + 800$ Hz (2500 Hz) and would be equal to 1600 Hz. The purpose of the bandpass filter is to limit the transmitted signal to this range. With 8-PSK, since one signaling element is equal to the time of 3 input bits, the output baud is equal to one-third the input bit rate or 1600 baud. Since 4800-bps transmission rate is accomplished using a bandwidth of 1600 Hz, the bandwidth efficiency is 4800 bps/1600 Hz or 3 bps/Hz. Note that the bandwidth range, 900 to 2500 Hz, is well within the 300- to 3000-Hz limitation imposed by the Telco facility. Three input bits are required to produce one output phase. If each phasor is considered a symbol, the required bandwidth of the BPF $(f_N) = f_s$, where $f_s = 4800$ bps/3 $= 1600$ Hz. This simply confirms a previous conclusion.

Figure 7–26   2- to 4-level converter outputs.

Quadrature Amplitude Modulation

In quadrature amplitude modulation (QAM), the amplitude and the phase of the transmitted signal are varied. 8-QAM can be accomplished by the modulator in Figure 7–27. In this modulator, a and b still determine the polarity of the converter output; c, however, is fed directly to both converters. If $c = 1$, both converters will output the higher magnitude; if $c = 0$, both converters will output the lower magnitude. With 8-PSK the output magnitudes of the two converters are never the same, while with 8-QAM the output magnitudes must always be the same. The *constellation diagram* of Figure 7–28 shows the possible output amplitudes and phases.

With 8-QAM, two amplitudes and four phases are possible. When performance is viewed in terms of vulnerability to noise, 16-QAM outperforms 16-PSK. For an average energy per bit-to-noise density ratio of 16 dB, 16-QAM has a probability of error of 10^{-8}, while 16-PSK has a probability of error of 10^{-4}. Therefore, when greater bandwidth efficiency is desired, QAM modems are used instead of PSK modems. A

Figure 7–27   8-QAM modulator.

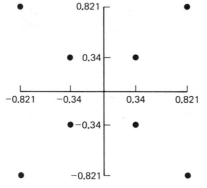

Figure 7–28   Constellation diagram of 8-QAM.

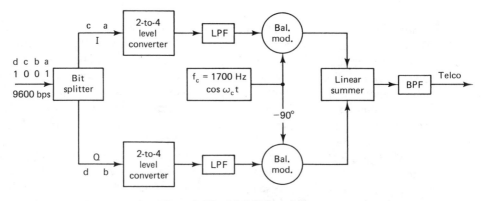

Figure 7–29   16-QAM modulator.

16-QAM modulator is shown in Figure 7–29. The bit splitter divides the incoming bit stream into two channels. The bit rate in each channel is 4800 bps. The 2- to 4-level converters require 2 bits before they produce one of four possible output levels. Bits a and b determine the polarity of the output, while bits c and d determine the magnitude.

$a, b = 0$: negative output $c, d = 0$: magnitude $= 0.22$ V

$a, b = 1$: positive output $c, d = 1$: magnitude $= 0.821$ V

Since the inputs to the two converters are totally independent of each other, each converter may output ± 0.22 V or ± 0.821 V. The purpose of the low-pass filter (LPF) is to eliminate all frequencies above the highest fundamental frequency at the converter output. If each level at the converter output is considered a symbol, 2 input bits to the converter will be required to produce one symbol at the output.

$$\frac{f_b}{2} = f_s$$

$$\frac{4800 \text{ bps}}{2} = 2400 \text{ symbols/second}$$

$$f_N \text{(LPF)} = \frac{f_s}{2} = \frac{2400}{2} = 1200 \text{ Hz}$$

This cutoff frequency for the LPF (f_N) is determined by observing the fastest rate of change at the converter output. This is shown in Figure 7–30. Since 2 input bits are required to produce one output level, 4 bits are required to produce one output square wave. The fundamental frequency of the square wave is one-fourth of the input bit rate or $4800/4 = 1200$ Hz. When all of the frequencies at the output of the LPF amplitude modulate the carrier, they produce frequencies ranging from 500 Hz ($f_c - f_N = 1700 - 1200 = 500$ Hz) to 2900 Hz ($f_c + f_N = 1700 + 1200 = 2900$ Hz) or a bandwidth of 2400 Hz. Since a 9600-bps transmission rate requires 2400 Hz of BW,

$$\text{BW efficiency} = \frac{9600 \text{ bps}}{2400 \text{ Hz}} = 4 \text{ bps/Hz}$$

The time of one signaling element is equal to the time of 4 bits. If 9600 bps is the input bit rate to the QAM modulator, the output baud is 2400. The phasor diagram for 16-QAM is shown in Figure 7–31. In this figure

Time of 2 bits

Time of 4 bits = time of one cycle

Figure 7–30 2- to 4-level converter output—QAM.

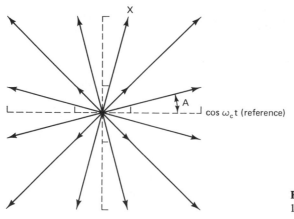

Figure 7–31   Phasor diagram for 16-QAM.

$$\text{angle } A = \tan^{-1}\frac{0.22}{0.821} = 15°$$

Therefore, the 16 possible outputs of this modulator include three different amplitudes and 12 different phases which are equally spaced 30° apart.

Example

For the input bit combinations shown in Figure 7–29:

Output of the I-channel converter: $+0.22$ V
Output of the Q-channel converter: -0.821 V
Output of the I-channel modulator: $+0.22 \cos \omega_c t$
Output of the Q-channel modulator: $-0.821 \sin \omega_c t$
Output of the BPF: $+0.22 \cos \omega_c t - 0.821 \sin \omega_c t$

This output phase is identified by X in Figure 7–31.

M-ary.   The term *M-ary* is often used in conjunction with PSK and QAM to identify the number of bits required to generate one output symbol.

$$n = \log_2 M$$

where n = number of bits
          M = number of combinations possible with n bits

M-ary can be used to make general statements concerning PSK or QAM. *M*-ary is derived from bi-nary, implying two different levels or conditions.

Example

PSK *M*-ary with $M = 8$ implies eight different phases are possible at the output. This has the same meaning as 8-PSK. To achieve eight possible phases, the input bit stream must be divided into tribits or groups of three.

Carrier recovery. If FSK is used, it can be seen from Figure 7–5 that the carrier need not be recovered by the receive modem in order to demodulate the receive signal. This is not the case with PSK or QAM. Not only must the exact carrier frequency be recovered, but its phase, when reinserted in the receive balanced modulator, must be identical to that at the transmit balanced modulator. *Coherent detection* implies that demodulation is accomplished using a reinserted carrier that has the *same frequency and phase* as the transmit carrier.

The input to the BPSK demodulator of Figure 7–10 is either $+\cos \omega_c t$ or $-\cos \omega_c t$. If either of these inputs is applied to a square-law device such as a FET, the square of either input, $\cos^2 \omega_c t = \frac{1}{2}(1 + \cos 2\omega_c t)$, will be generated. The double-frequency component is filtered out and passed through a frequency-divider circuit to recover the frequency and phase of the transmitted carrier. With 4-PSK, 8-PSK, and 16-QAM, carrier frequency and phase recovery are much more complex and expensive to implement.

Differential Phase Shift Keying

As an alternative, differential phase detection is preferred. Rather than compare the phase of the received signal to a fixed reference, the phase of the present received signal is compared to the phase of the previously received signal. Differential phase detection necessitates differential phase encoding. Although differential phase detection is easier to implement, it has associated disadvantages. The signal-to-noise ratio must be 1 to 3 dB greater using differential detection to achieve the same error rate as coherent detection. A DBPSK modulator is shown in Figure 7–32. The underlined bit in the XNOR output is a reference bit. Note that the bit that modulates the carrier depends not only on the value of the current bit but also on the value of the bit immediately before it. The received signal will be $+\sin \omega_c t$ or $-\sin \omega_c t$. Since the balanced modulator in

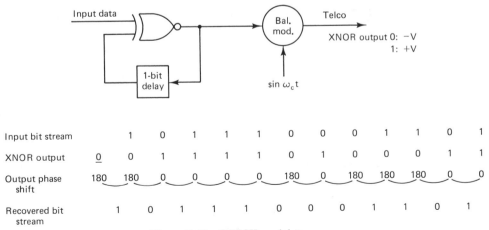

Input bit stream		1	0	1	1	1	0	0	0	1	1	0	1
XNOR output	<u>0</u>	0	1	1	1	1	0	1	0	0	0	1	1
Output phase shift	180	180	0	0	0	0	180	0	180	180	180	0	0
Recovered bit stream		1	0	1	1	1	0	0	0	1	1	0	1

Figure 7–32 DBPSK modulator.

Figure 7–33 produces the product of two successive inputs, its output will be $\sin^2\omega_c t = \frac{1}{2}(1 - \cos 2\omega_c t)$ when two consecutive phases are the same, or $-\sin^2\omega_c t = -\frac{1}{2}(1 - \cos 2\omega_c t)$ when the two consecutive phases are different. The LPF eliminates the double-frequency component and yields either a positive or a negative dc level which represents a logic 1 or 0, respectively.

Figure 7–33 DBPSK demodulator.

MODEM SYNCHRONIZATION

During the RTS/CTS delay, the transmit modem outputs a special, internally generated bit pattern called the *training sequence* (Chapter 6). This bit pattern is used to synchronize (train) the receive modem. Depending on the type of modulation, transmission bit rate, and the complexity of the modem, the training sequence accomplishes one or more of the following functions in the receive modem:

1. Verify continuity (activate RLSD).
2. Initialize the descrambler circuits. (These circuits are used for clock recovery—explained later in this chapter.)
3. Initialize the automatic equalizer. (These circuits compensate for telephone line impairments—explained later in this chapter.)
4. Synchronize the transmitter and receiver carrier oscillators.
5. Synchronize the transmitter and receiver clock oscillators.
6. Disable any echo suppressors in the circuit.
7. Establish the gain of any AGC amplifiers in the circuit.

Low-Speed Modems

Since these modems are generally asynchronous (Table 7–1) and use noncoherent FSK, the transmit carrier and clock frequencies need not be recovered by the receive modem. Therefore, scrambler and descrambler circuits are unnecessary. The pre- and post-equalization circuits, if used, are generally manual and do not require initialization. The special bit pattern transmitted during the RTS/CTS delay is usually a constant string of 1's (idle line 1's) and is used to verify continuity, set the gain of the AGC amplifiers, and disable any echo suppressors in dial-up applications.

Medium- and High-Speed Modems

These modems are used where transmission rates of 2400 bps or more are required. In order to transmit at these higher bit rates, PSK or QAM modulation is used which requires the receive carrier oscillators to be at least frequency coherent (and possibly phase coherent). Since these modems are synchronous (Table 7–1), clock timing recovery by the receive modem must be achieved. These modems contain *scrambler* and *descrambler circuits* and *adaptive (automatic) equalizers.*

Training. The type of modulation and encoding technique used determines the number of bits required and therefore the duration of the training sequence. The Bell System 208 modem is a synchronous, 4800-bps modem which uses 8-DPSK. The training sequence for this modem is shown in Figure 7–34. Each symbol represents 3 bits (1 tribit) and is 0.625 ms in duration. The four-phase idle code sequences through four of the eight possible phase shifts. This allows the receiver to recover the carrier and the clock timing information rapidly. The four-phase test word allows the adaptive equalizer in the receive modem to adjust to its final setting. The eight-phase initialization period prepares the descrambler circuits for eight-phase operation. The entire training sequence (234 bits) requires 48.75 ms for transmission.

Figure 7–34   Training sequence for a 208 modem.

Clock recovery.   Although timing (clock) synchronization is first established during the training sequence, it must be maintained for the duration of the transmission. The clocking information can be extracted from either the I or the Q channel, or from the output of the bit combiner (Figure 7–19). If an alternating 1/0 pattern is assumed at the output of the LPF (Figure 7–35), a clock frequency at the bit rate of the I (or Q) channel can be recovered. The waveforms associated with Figure 7–35 are shown in Figure 7–36.

This clocking information is used to phase-lock loop the receive clock oscillator onto the transmitter clock frequency. To recover clocking information by this method successfully, there must be sufficient transitions in the received data stream. That these transitions will automatically occur cannot be assumed. In a QPSK system, an alternating 1/0 pattern applied to the transmit modulator produces a sequence of all 1's in the I or Q channel, and a sequence of all 0's in the opposite channel. A prolonged sequence of all 1's or all 0's applied to the transmit modulator would not provide any transitions in either the I, Q, or the composite received data stream. Restrictions could

Figure 7–35 Clock recovery circuit for QPSK demodulator.

Figure 7–36 Clock recovery from I (or Q) channel of a QPSK demodulator.

be placed on the customer's protocol and message format to prevent an undesirable bit sequence from occurring, but this is a poor solution to the problem.

Scramblers and descramblers. A better method is to scramble the customer's data before it modulates the carrier. The receiver circuitry must contain the corresponding descrambling algorithm to recover the original bit sequence before data are sent to the DTE. The purpose of a scrambler is not simply to randomize the transmitted bit sequence, but to detect the occurrence of an undesirable bit sequence and convert it to a more acceptable pattern.

A block diagram of a scrambler and descrambler circuit is shown in Figure 7–37. These circuits are incomplete since an additional gate would be required to detect a varying sequence that would create an all 1 or all 0 sequence in a modulator channel after the bits were split.

Example

For QPSK or 4-QAM: 0 1 0 1 0 1 0 1 0 1 0 1
                            I Q I Q I Q I Q I Q I Q

Figure 7–37 Scrambler and descrambler circuits.

For 8-PSK or 8-QAM:

0 1 1 0 1 0 0 1 1 0 1 1
I Q   I Q   I Q   I Q

The scrambler circuit is inserted prior to the bit splitter of the QPSK modulator of Figure 7–17, and the descrambler is inserted after the bit combiner of the QPSK demodulator in Figure 7–19. In general, the output of the scrambler or descrambler OR gate is A B C D + A′ B′ C′ D′.

$$f_m = f_d \oplus (\text{A B C D} + \text{A}' \text{ B}' \text{ C}' \text{ D}') \qquad \text{top XOR gate}$$

$$f_r = f_m \oplus (\text{A B C D} + \text{A}' \text{ B}' \text{ C}' \text{ D}') \qquad \text{bottom XOR gate}$$

Substituting for f_m in the second equation, we have

$$f_r = f_d \oplus (\text{A B C D} + \text{A}' \text{ B}' \text{ C}' \text{ D}') \oplus$$
$$(\text{A B C D} + \text{A}' \text{ B}' \text{ C}' \text{ D}')$$

Since any identity XORed with itself yields 0,

$$f_r = f_d \oplus 0$$
$$f_r = f_d$$

This simply shows that the original transmitted data (f_d) will be fully recovered by the receiver.

The output of either OR gate will be a 1 if the 4-bit register contains either all 1's or all 0's. Neither of these is a desirable sequence. If the OR gate output is a 1, f_m will be the complement (opposite) of f_d, or f_r will be the complement of f_m. The intent is to create transitions in a prolonged bit stream of either all 1's or all 0's. If the output of the OR gate is a 0, neither of these undesired conditions exists and $f_m = f_d$ or $f_r = f_m$: the data pass through the XOR gate unchanged. If the other logic gates (AND, OR, NAND, NOR) were used either alone or in combination in place of the XOR gates, the necessary transitions could be created in the scrambler circuit, but the original data could not be recovered in the descrambler circuit. If a long string of all 1's or all 0's is applied to the scrambler circuit, this circuit will introduce transitions. However, there may be times when the scrambler creates an undesired sequence. The XOR output is always either a 1 or a 0. No matter what the output of the OR gate, a value of f_d may be found to produce a 1 or a 0 at the XOR output. If either value for f_d was equiprobable, the scrambler circuit would be unnecessary. If the 4-bit register contains all 1's, if $f_d = 1$, we would like to see it inverted. However, if $f_d = 0$, we'd prefer to pass it through the XOR gate unchanged. The scrambler circuit for this situation inverts the 0 and extends the output string of 1's. It is beyond the intended scope of this book to delve deeply into all parameters involved in scrambler design. Let it be enough to say that scramblers will cure more problems than they create.

Equalizers.   *Equalization* is the compensation for the phase delay distortion and amplitude distortion of a telephone line. One form of equalization is C-type conditioning (Chapter 3). Additional equalization may be performed by the modems. *Compromise equalizers* are contained in the transmit section of the modem and they provide *pre-equalization.* They shape the transmitted signal by altering its delay and gain characteristics before it reaches the telephone line. It is an attempt to compensate for impairments anticipated in the bandwidth parameters of the line. When a modem is installed, the compromise equalizer is manually adjusted to provide the best *bit error rate* (BER). Typically, compromise equalizer settings affect:

1. Amplitude only
2. Delay only
3. Amplitude and delay
4. Neither amplitude nor delay

The setting above may be applied to either the high or low voice band frequencies or symmetrically to both at the same time. Once a compromise equalizer setting has been selected, it can only be changed manually. The setting that achieves the best BER is dependent on the electrical length of the circuit and the type of facilities that make it up. *Adaptive equalizers* are located in the receiver section of the modem and provide *post-equalization* to the received analog signal. Adaptive equalizers automatically adjust their gain and delay characteristics to compensate for telephone-line impairments. An adaptive equalizer may determine the quality of the received signal within its own circuitry or it may acquire this information from the demodulator or descrambler circuits. Whichever the case, the adaptive equalizer may continuously vary its settings to achieve the best overall bandwidth characteristics for the circuit.

SIGNAL QUALITY

A common method of determining the signal quality is to sample the power within discrete passbands in the receive signal and compare the power distribution to a known reference. The spectral distribution of the reference is based on the spectral content of a typical transmitted signal at the output of the modulator. Then, if the received distribution does not resemble a typical transmitted signal, the equalizer automatically adjusts its passband characteristics until the best received signal quality is achieved.

The quality of the received signal can also be determined by monitoring the received data entering the descrambler circuit. The receive modem cannot determine what the bit sequence should be, but it can recognize bit sequences that the transmitter scrambler circuit prohibits from occurring under normal circumstances. Typically, prolonged bit sequences of consecutive 1's and 0's are not allowed to modulate the carrier. If one of these prohibited bit sequences is detected, the equalizer concludes that transmission errors must be occurring. The equalizer then automatically adjusts its passband characteristics in an attempt to correct the suspected problem. Very often, the front panel of a modem has an LED indicator to identify when the adaptive equalizer has sequenced through all of its tab settings and is still receiving marginal data. High-speed (synchronous) modems are usually equipped with automatic retrain circuitry. The first attempt by an adaptive equalizer to improve the quality of the received signal is accomplished with customer data or idle line 1's. If unsuccessful, the receive modem initiates a request for training by sending out its own training sequence. Whenever a modem receives a training sequence while it is transmitting, it will, in turn, suspend data transmission and send its training sequence. This is a destructive process and interrupts data transmission. Only modems capable of transmission rates of 9600 bps are equipped with automatic retraining circuitry, and this circuitry is used only if the modems were optioned for continuous carrier operation. This necessarily implies a two-point, four-wire, FDX system.

In summary, a block diagram of a synchronous PSK modem is shown in Figure 7–38. A brief functional description of each block follows.

Transmitter

Transmitter timing. This circuit generates the various clock signals required by the modem transmitter. The master clock is either internally generated or derived from the signals on SCTE from the DTE.

Start sequence controller. When RTS goes high, this circuit enables the transmission of the analog carrier through switch 1. The carrier is still unmodulated by

Figure 7-38 Block diagram of a synchronous PSK modem.

terminal data because switch 2 is disabled. When RTS goes high, this circuit also signals the pattern generator/scrambler to initiate the transmission of the training sequence. This controller provides the RTS/CTS delay, activating switch 2 and placing a high on CTS after this time lapse. When RTS goes low, the start sequence controller deactivates switch 1, removing the carrier from the telephone line.

Pattern generator and scrambler. The pattern generator provides the special bit pattern transmitted during the training sequence, and the idle line 1's for continuous carrier applications while there is no transmission of data. The scrambler randomizes the input data to provide the transitions necessary for clock recovery.

Phase modulator. This circuit converts binary input data from the DTE to corresponding phase changes in the analog carrier. The phase modulator uses two binary channels to achieve differential PSK at its output.

Compromise equalizer. This circuit is a manually adjustable filter which provides pre-equalization to the voice band signals. This equalizer compensates for anticipated impairments in the gain and delay characteristics introduced by the telephone circuit.

Amplifier or pad. This is a variable-gain amplifier capable of inserting loss or gain into the transmitted signal path. It is used to set the transmit signal power during initial installation.

T, R, T1, R1. Transmit and receive tip and ring.

The input and output amplifiers are the drivers and terminators of the RS 232C interface.

Receiver

Adaptive equalizer. This circuit provides post-equalization of the line's gain and delay characteristics.

Signal quality monitor. This circuit monitors the received analog signal and determines whether the adaptive equalizer is properly set to compensate for the telephone line impairments. If not, the equalizer startup controller is signaled to retrain locally the adaptive equalizer. During the retraining time, the received data may be invalid.

Equalizer startup controller. This circuit adjusts the adaptive equalizer when this modem is receiving a training sequence or when the signal quality of the received signal becomes substandard.

Carrier recovery. This circuit detects the presence of a received carrier and controls the on/off condition of the RLSD line of the RS 232C interface. This circuit also recovers the carrier and phase-lock loops the local oscillator to this frequency to produce a frequency coherent signal for demodulation.

Demodulator. This circuit converts the phase variations of the received carrier to the corresponding bit sequence.

Clock recovery. This circuit recovers the transmitted clock frequency and generates the various clocking signals required by the receiver.

Descrambler. After the clock has been recovered, this circuit converts the randomized data back to the original transmitted data.

DIAL-UP MODEMS

Dial-up modems must satisfy different requirements than those designed for private-line usage. Access into and out of the DDD network must be made in a manner similar to that of a normal telephone connection. The subscriber must make the modem look and act like a telephone set to the network. There are two general methods of coupling the modem into and out of the DDD network, *hardwire* and *acoustic coupling.* Hardwire coupling is accomplished through a data coupler, while acoustic coupling permits the convenience of using the telephone set and the network as it stands.

Message Transfer through an Acoustical Coupler

The operator at the originate end of the circuit initiates the call by picking up the telephone handset. Generally, the telephone is a standard 500-type telset [Figure 7–39(a)]. When the telset is picked up, an off-hook condition is applied to the telephone line. The Telco network responds with an audible dial tone, indicating to the operator that a successful access to the DDD network has been made. The operator then dials the telephone number associated with the answer modem. At the answer end, when an incoming ring is detected, the ring indicator (pin 22) of the RS 232C interface goes active. If the associated DTE equipment (CPU) is prepared and capable of receiving a message, it allows the data terminal ready pin (pin 20) of the RS 232C interface to go active, signaling the modem to answer the call and transmit the F2 mark frequency. On

Handset

(a)   (b)

Figure 7–39 (a) Telephone; (b) modem.

hearing this tone, the operator at the originate end places the handset in the cradle of the acoustically coupled modem [Figure 7–39(b)]. The acoustic input (F2 mark frequency) is converted to electrical energy by the modem microphone (see Figure 7–40).

The microphone allows this signal to pass through A2 to the carrier detect circuit. A2 is activated by either a mechanical switch or a photocell in the telset cradle. A2 prevents any electrical signal from entering the receive section of the modem unless the handset is in the cradle. The bandpass filter allows only the F2 mark/space frequencies to pass. When the originate modem detects the F2 mark frequency, it signals the DTE by making RLSD (pin 8) of the RS 232C interface active. After the RTS/CTS delay, F1 mark is transmitted by the originate modem. Switch A1 assures that the F1 mark is not transmitted without a carrier detect signal (without having access to the DDD network). The originate modem is now in the data mode and data transfer begins and continues until either of the DTEs breaks the carrier and terminates the call. Signaling between the modems and their corresponding DTEs is accomplished either through the RS 232C interface, a 20-mA loop, or some other serial interface.

In the transmission of data, digital pulses are sent from the line control unit (LCU) to the FSK modulator through the RS 232C interface. These pulses modulate the VCO and produce analog changes at its output. The transmitter bandpass filter allows only the F1 mark/space frequencies to pass. The speaker converts these analog signals to

Figure 7–40 Block diagram of a 103-compatible acoustic coupled modem. The AND gates here do not necessarily perform a logic AND function on the inputs. They simply indicate that both inputs must be present before an output can be obtained.

audible tones which enter the microphone in the handset and are transmitted over the telephone lines in the same manner as voice.

Echo-Plex

Echo-plex is often used with low-speed acoustically coupled modems. It is a mode of transmission that achieves less than full duplex but more than half-duplex. Inexpensive, less sophisticated terminal equipment that is commonly used with acoustically coupled modems generally do not provide an electrical connection from the keyboard to the CRT. When a key is depressed on the keyboard, the corresponding character is not sent directly to the CRT for display. In Figure 7–40, with S1 in the HDX position, a transmitted character is looped back to the DTE receiver locally through the wave-shaping circuit. The receiver decodes and displays that character on the CRT. The modem is limited to half-duplex operation: if data were being received from the FSK demodulator at the same time the DTE was transmitting, the two signals would become garbled in the wave-shaping circuit. By placing S1 in the FDX position, the transmitted data cannot reach the wave-shaping circuit directly. This prevents the transmitted character from being displayed on the CRT. Echo-plex is achieved by having the answer DTE retransmit (echo) the received message back to the originate DTE for decoding and display. Although transmission occurs in both directions simultaneously as in FDX, only one terminal can transmit its message at any given time as in HDX. The advantage of echo-plex is that the operator at the originate DTE, when he or she sees a character displayed on his CRT, knows that it has been received by the answer terminal.

Message Transfer through Hardwire Attachment

The modem can access the DDD network through a hardwire coupler called a *data access arrangement (DAA)*. An *automatic data access arrangement (ADAA)* enables the terminal equipment to originate a call automatically (give off-hook supervision), dial into the network (provide pulses or touch-tone signals), answer a call (respond to an incoming ring), and terminate a call (give on-hook supervision). The Bell System provides two such couplers, CBT and CBS. *CBS* is designed to conform to RS 232C electrical standards, while *CBT* is designed to provide contact closures for less complex electromechanical communications equipment.

The major functions of the CBS and CBT data couplers are:

1. To provide a transmission path from the customer-provided data modems to the DDD network
2. To protect Telco personnel and equipment from hazardous voltages which may be accidentally supplied by the customer-provided modems
3. To limit the signal power from the data modem to a specified value in order to prevent interference with other telephone services
4. To protect the telephone line from longitudinal imbalance (producing currents in the same direction on a two-wire pair)

5. To provide the following functions associated with network control signaling:
   (a) To provide a loop-holding path for dc supervision
   (b) To detect ringing and alert the customer's terminal to an incoming call
   (c) To originate on-hook and off-hook signaling and to generate dial pulses in response to signals received from the customer through the interface control leads
   (d) To provide a delay of 1 to 3 s after an incoming call is answered in order to prevent data signals from interfering with automatic message accounting equipment
6. To provide the capability of remote testing the data coupler
7. To provide an indication of the status of the switch hook of the associated telephone

A block diagram of a CBS data coupler is shown in Figure 7–41. The tip and ring (T, R) pins correspond to the tip and ring of a two-wire telephone facility. The incoming 20-Hz ring signal is rectified and integrated by the ring detector until there is sufficient buildup of charge to ensure the DAA that the ring is legitimate. This prevents normal line transients, produced by switching and dialing, from supplying false ring indications. The coupler supplies a dc voltage on the RI pin to notify the modem of an incoming ring. The modem responds with an ON signal on the OH (off-hook) pin. This signal actuates the OH relay, allowing a dc current path for the two-wire telephone loop. The Telco dial

Figure 7–41 Simplified block diagram of a data access arrangement.

switch recognizes this as an off-hook condition and stops the ringing signal. The modem now applies an ON signal to the DA pin, which actuates the CT relay. Actuating this relay removes the resistive load from the automatic level control in the modem and allows the T and R leads to be coupled (cut through—CT) to the DT and DR (data tip and data ring) leads. Activation of the CT relay also produces an ON signal on the CCT pin, indicating to the modem that the Telco facilities are coupled to the modem. The modem now waits for the carrier to be sent. A call is terminated when the modem supplies a low to the OH pin which disables the OH relay and opens the dc path to the dial switch. A call is originated in much the same manner. The modem applies an ON signal to the OH pin to complete a dc path to the dial switch. The OH signal is now pulsed to represent dial pulses to the telephone switching equipment. After the dial pulses have been completed, the DA pin is made high to activate the CT relay and cut the Telco facilities through to the modem. The modem, on receipt of an ON signal on CCT, will begin message transmission. After the message has been sent, the modem allows OH to go low, terminating the Telco Connection.

QUESTIONS

1. A modem is a form of analog-to-digital/digital-to-analog converter. (T, F)
2. If, when you made a phone call, your voice frequencies were in analog form end to end, would modems be required for transmission or reception?
3. What is meant by a synchronous modem?
4. Asynchronous modems do not require a clock input. (T, F)
5. What is the fundamental frequency of a 1200-Hz square wave?
6. What is the maximum data transmission rate of modems using:
   (a) FSK?   **(b)** PSK?   **(c)** QAM?
7. FSK can be considered a form of FM. (T, F)
8. A backward or reverse channel is normally provided to modems designed to operate on a four-wire system. (T, F)
9. The reverse channel uses FSK modulation. (T, F)
10. The reverse channel is used mainly for data link control signals. (T, F)
11. Modems with originate/answer options are normally used in full-duplex systems. (T, F)
12. With originate/answer modems, F2 refers to the originate mark/space frequencies. (T, F)
13. As the transmission rate increases:
    (a) The h factor increases. (T, F)
    (b) The number of significant frequencies present in the transmitted signal increases. (T, F)
    (c) The separation between the frequencies contained in the transmitted signal increases. (T, F)
    (d) The bandwidth required to transmit all of the significant frequencies increases. (T, F)
14. The value of J_0 on the Bessel function chart indicates the relative amplitude of the first pair of side frequencies contained in the transmitted waveform. (T, F)
15. If dial-up lines are used for FDX operation, some form of multiplexing must be accomplished. (T, F)

16. With FSK, bandwidth limitations restrict the maximum bit transmission rate. (T, F)
17. The carrier frequency is present at the output of a balanced modulator. (T, F)
18. In BPSK, what is the function of the low-pass filter (Figure 7–10)?
19. By increasing the bandwidth efficiency, the data transmission rate can be increased without an accompanying increase in bandwidth requirements. (T, F)
20. Baud is always synonymous with bits per second. (T, F)
21. The output baud can be greater than the input bit rate. (T, F)
22. The output baud can be less than the input bit rate. (T, F)
23. Tribits are associated with (*BPSK, QPSK, 8-PSK, QAM*).
24. In an *M*-ary system with $M = 16$:
    (a) How many different output conditions are possible?
    (b) How many bits are required to determine each of these conditions?
25. DPSK uses coherent phase detection. (T, F)
26. Identify one advantage of DPSK over PSK.
27. Identify one disadvantage of DPSK over PSK.
28. A 4800-bps data stream is applied to a:
    (a) QPSK modem   **(b)** 8-PSK modem   **(c)** QAM modem
    Answer the following questions for each of the three modems.
    (1) How many bits are considered before the output phase and amplitude are determined?
    (2) How many different output phases are possible?
    (3) Ideally, what is the degree separation between phasors?
    (4) How many different amplitudes can the output have?
    (5) What is the output baud?
29. If 8-DPSK was used and the following sequence of phases was detected:

$$..., +45, +90, +135, +45$$
$$\quad\quad A \quad\quad\quad\quad\quad\quad\quad B$$

Phasor B would be (*leading, lagging, in phase with, 180° out of phase with*) phasor A. By how many degrees?
30. What is meant by "coherent frequency detection" and "coherent phase detection"?
31. The bit pattern in the training sequence for high-speed modems is a continuous sequence of 1's. (T, F)
32. Identify all possible locations where clocking information can be extracted in a PSK or a QAM demodulator?
33. What is the purpose of a scrambler circuit?
34. What does equalization provide?
35. What is the difference between a compromise equalizer and an adaptive equalizer?
36. In reply to a transmitted message, a modem receives a training sequence. This high-speed modem will (*retransmit the message, accept the training sequence as a positive acknowledgment and continue with message transmission, suspect internal malfunctions and train onto the training sequence, send out its own training sequence*).
37. Only high-speed modems are equipped with automatic retraining circuitry. (T, F)

38. With dial-up modems, how does the operator at the originate end know when the circuit has been completed to the receive modem?

39. A dial-up modem cannot transmit unless it is detecting a receive carrier signal.   (T, F)

40. How does FDX in echo-plex operation differ from normal FDX?

PROBLEMS

1. $f_c = \sin \omega_c t$ at 1500 Hz; input bit rate = 2000 bps; $b,\ a = 0,\ 0$.

2. $f_c = \cos \omega_c t$ at 1550 Hz; input bit rate = 1800 bps; $b,\ a = 1,\ 0$.

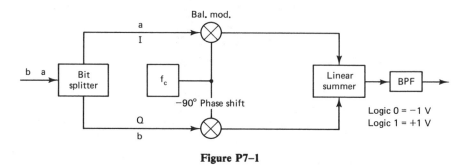

Figure P7–1

With reference to Figure P7–1 and for the data given in Problems 1 and 2, answer the following questions:

(a) What is the bit rate in the I channel?

(b) What is the fundamental frequency of the highest-frequency square wave applied to the balanced modulator?

(c) What is the trigonometric expression for the output of the BPF for the input bits given ($\omega_c t \pm 90°$ to be eliminated from the output expression)?

(d) What is the bandwidth at the output of the BPF?

(e) The bandwidth of part (d) extends from _____ to _____ Hz.

(f) What is the output baud?

(g) Identify the values of input bits *a* and *b* that will provide the output phasors shown. Use the phase of f_c as reference.

Figure P7–1g

(h) For the corresponding demodulator circuit (Figure 7–19), show trigonometrically how the original data bits are recovered.

3. $f_c = \sin \omega_c t$ at 1600 Hz; phase shift $= -90°$; $c,\ b,\ a = 1,\ 1,\ 0$; input bit rate = 4500 bps.

4. $f_c = \cos \omega_c t$ at 1500 Hz; phase shift $= +90°$; c, b, $a = 0, 1, 1$; input bit rate $= 4200$ bps. With reference to Figure P7–3 and for the data given in Problems 3 and 4, answer the following questions:

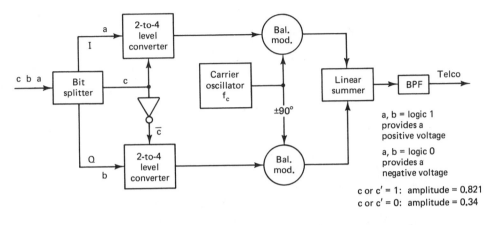

Figure P7–3

- **(a)** What is the bit rate of the I channel?
- **(b)** What is the fundamental frequency of the highest-frequency square wave applied to the balanced modulator?
- **(c)** What is the trigonometric expression for the output of the BPF for the input bits given ($\omega_c t \pm 90°$ to be eliminated from the output expression)?
- **(d)** What is the bandwidth at the output of the BPF?
- **(e)** The bandwidth of part (d) extends from _____ to _____ .
- **(f)** What is the output baud?
- **(g)** What is the bandwidth efficiency?

5. $f_c = \sin \omega_c t$ at 1500 Hz; input bit rate $= 9200$ bps; d, c, b, $a = 1, 1, 0, 0$; phase shift $= +90°$.

6. $f_c = \cos \omega_c t$ at 1600 Hz; input bit rate $= 8800$ bps; d, c, b, $a = 0, 1, 1, 0$; phase shift $= -90°$.

In Figure P7–5, the first bit into the 2- to 4-level converter determines the sign: $0 = -$; $1 = +$. The second bit into the 2- to 4-level converter determines the amplitude: for Problem 5, $0 = 0.6$, $1 = 0.9$; for Problem 6, $0 = 5$, $1 = 8.66$. Using this information, together with the data given for Problems 5 and 6, answer the following questions:

- **(a)** What is the bit rate into the 2- to 4-level converter?
- **(b)** How many input bits into the 2- to 4-level converter are required to obtain one symbol (one of the four levels) out?
- **(c)** What is the symbol rate into the LPF?
- **(d)** Each symbol can have four possible values. What are they?
- **(e)** What is the f_N for the LPF?
- **(f)** What is the double-sided f_N for the BPF?
- **(g)** The bandwidth of the BPF extends from _____ to _____ .
- **(h)** What is the output baud?

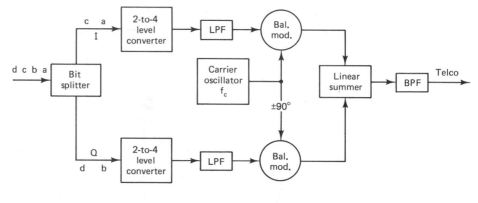

Figure P7–5

(i) What is the BW efficiency?

(j) For the given input bits, what is the trigonometric expression for the output of the BPF?

(k) What are the three possible amplitude values at the BPF output?

(l) The degrees separating two successive phasors is either _____ or _____ .

7. For the DBPSK modulator in Figure 7–32, for the given input bit stream, identify the binary output sequence of the XNOR gate with the provided reference bit. The order of the arrow is to the left.

	Reference Bit	Input Bit Sequence
(a)	1	1 1 0 0 1 0 1 1 0
(b)	0	1 1 0 0 1 0 1 1 0

8. For the DBPSK demodulator in Figure 7–33, identify the output bit sequence for the phase inputs identified and the reference phase given.

	Reference Phase	Phase Inputs
(a)	$0°$	$0°$ $180°$ $180°$ $180°$ $0°$ $180°$ $0°$ $180°$ $0°$
(b)	$0°$	$180°$ $180°$ $0°$ $180°$ $0°$ $180°$ $180°$ $180°$ $180°$

8

VOICE OR DATA TRANSMISSION WITH ANALOG CARRIERS

FREQUENCY-DIVISION MULTIPLEXING

Multiplexing is the transmission of information from more than one source on the same media. In *frequency-division multiplexing* (FDM), many information channels are transmitted simultaneously, with each channel occupying a different frequency band (Figure 8–1). If each information channel originally occupied the same frequency range, the frequencies must be translated to different areas of the frequency spectrum before they are combined. To achieve frequency separation, each channel amplitude-modulates a different carrier frequency. If a carrier is amplitude-modulated with a single frequency, the resultant waveform is mathematically described as

$$A \sin \omega_c t + \frac{\text{mA}}{2} \cos (\omega_c - \omega_m)t - \frac{\text{mA}}{2} \cos (\omega_c + \omega_m)t$$

$$(a) \qquad\qquad (b) \qquad\qquad (c)$$

where A = peak carrier amplitude
m = modulation coefficient
$\omega_c = 2\pi f_c$
$\omega_m = 2\pi f_m$
f_c = carrier frequency
f_m = modulating frequency

Expression (a) is the original carrier frequency, (b) the lower side or difference frequency, and (c) the upper side or sum frequency.

If a carrier is amplitude-modulated by a band of frequencies, an upper and a lower sideband are produced. The *upper sideband* (USB) is made up of the sum of the carrier

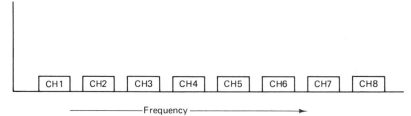

Figure 8–1   Frequency-division multiplexing.

frequency and the individual frequencies present in the modulating signal: the *lower sideband* (LSB) is made up of the difference between the carrier frequency and the individual frequencies present in the modulating signal (Figure 8–2).

In amplitude modulation the carrier contains no intelligence; therefore, it is suppressed through some form of balanced modulator (see "Ring Modulator" in Chapter 7). Since the upper and lower sidebands contain identical information, the transmission of only a single sideband is necessary to convey the information. With FDM, a single sideband is transmitted without the carrier. This signal is described as single-sideband suppressed carrier (SSBSC).

Figure 8–2   Frequency spectrum of an AM signal and a SSB-SC.

An A-type (analog) channel bank performs frequency-division multiplexing of twelve voice band channels. Each voice band channel can carry either voice information or digital information from a modem. Each channel amplitude-modulates a different carrier frequency. The lower sideband of each modulation process is extracted and combined with the lower sidebands from the eleven other channels to form a *group* (Figure 8–3). A group has a bandwidth of 48 kHz (12 × 4 kHz) and occupies the frequency band from 60 to 108 kHz. Although each voice channel is allocated a frequency range of 0 to 4 kHz, signal information is normally limited to a 300- to 3000-Hz passband. Consequently, a group has a natural guard band of 1.3 kHz (Figure 8–4) between adjacent channel signals.

If further multiplexing is desired, five groups may be similarly combined to produce a *super group* (SG). The bandwidth of an SG, which results from combining the LSBs of each modulation process, is 240 kHz and extends from 312 to 552 kHz (Figure 8–5).

Super groups may be combined to form a *master group*. A master group is made

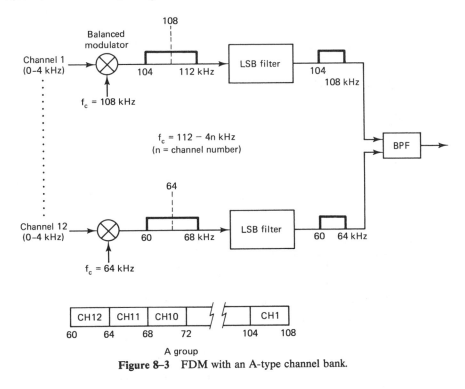

A group

Figure 8-3 FDM with an A-type channel bank.

Figure 8-4 Guard band between signals in FDM.

up of ten super groups and contains information from 600 voice band channels. A scheme that forms the U600 master group is shown in Figure 8–6. The guard band between the adjacent super groups in Figure 8–6 is 8 kHz and the guard band between adjacent master groups in Figure 8–7 is 80 kHz. This guard band makes it easier to separate adjacent super groups and master groups with filters in the FDM receiver. An L600 master group combines super groups 1 through 10 in a slightly different method and occupies the bandwidth from 60 to 2788 kHz. The L600 master groups are transmitted either directly on coaxial cable or are modulated further and then transmitted on microwave radio. U600 master groups may be transmitted as is or used as stepping stones for further multiplexing. In Figure 8–7, three master groups are multiplexed to form one *microwave radio channel.*

A super group may be added to the three multiplexed master groups (1860 VB channels) and transmitted on either coaxial cable or microwave radio. *Jumbo groups* (3600 VB channels) combine six master groups and are transmitted on coaxial cable. Three jumbo groups may be further multiplexed (10,800 VB channels) and transmitted on coaxial cable.

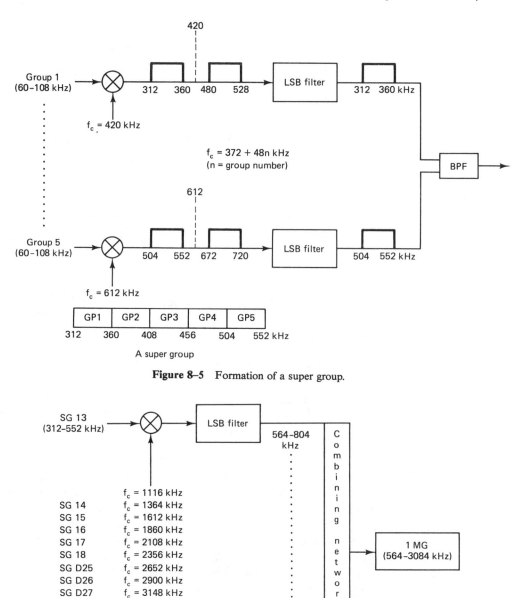

Figure 8–5 Formation of a super group.

Figure 8–6 U600 master group.

Generation of the Carrier Frequency

An FDM receiver accomplishes demodulation by successively mixing the baseband signal down in frequency until the original VB channel frequencies are recovered. In

Figure 8-7   Formation of a radio channel.

FDM systems, it is essential for the transmit and the receive carrier frequencies to be synchronized. If they are not, the recovered VB signals would be offset in frequency from their original spectrum by the difference in the two carrier frequencies. For the purpose of power conservation, balanced modulators, which suppress the carrier, are used in FDM transmitters. Therefore, the transmit carrier cannot be recovered at the receiver directly from the composite baseband signal. Note that all of the carrier frequencies in Figures 8-3 through 8-6 are multiples of 4 kHz. In a communications system, one station is designated as the master station. The 4-kHz master oscillator, to which all stations in the system are synchronized, is located at this station. Generally, the 4-kHz base frequency is multiplied to a higher pilot frequency (typically 64, 312, or 552 kHz) and combined with the composite baseband spectrum. Each secondary station demodulates the pilot, then recovers and regenerates the 4-kHz base frequency. Consequently, all stations in a system derive their carrier frequencies from coherent 4-kHz carrier supplies. In a large system, such as the Bell System or General Telephone, it is impractical for the master station to transmit the pilot directly to each slave station. Instead, many slave stations also serve as repeaters to other slave stations for the pilot.

The carrier frequencies used in FDM are obtained by passing the 4-kHz base frequency through a nonlinear device, then extracting the appropriate harmonics. If the 4-kHz oscillator drifts in frequency or shifts in phase, the harmonics will change proportionally.

Example

What is the radio channel output frequency range of a voice band channel belonging to channel 10, group 4, super group 17, master group 3?

	Ch 10 at GP	GP 4 at SG	SG 17 at MG	MG 3 at Radio Channel
f_c	72 kHz	564 kHz	2108 kHz	8848 kHz
LSB	68–72 kHz	492–496 kHz	1612–1616 kHz	7232–7236 kHz

From the example on page 123 it can be seen that as a voice band channel is elevated in frequency, it still occupies the original 4-kHz bandwidth.

Example

If the 4-kHz oscillator drifts 10 Hz, by how much would a 1-kHz tone on channel 3, group 2, super group 17, master group 2 differ from what it should be?

Assuming that all carrier frequencies that are amplitude-modulated are derived as harmonics of the 4-kHz signal:

For Channel 3
  Ideal f_c: 100 kHz
  Actual f_c: 4.01 kHz $\times$ 25 $=$ 100.25 kHz
  Ideal LSF: 100 kHz $-$ 1 kHz $=$ 99 kHz
  Actual LSF: 100.25 kHz $-$ 1 kHz $=$ 99.25 kHz

For Group 2
  Ideal f_c: 468 kHz
  Actual f_c: 4.01 kHz $\times$ 117 $=$ 469.17 kHz
  Ideal LSF: 468 kHz $-$ 99 kHz $=$ 369 kHz
  Actual LSF: 469.17 kHz $-$ 99.25 kHz $=$ 369.92 kHz

For Super Group 17
  Ideal f_c: 2108 kHz
  Actual f_c: 4.01 kHz $\times$ 527 $=$ 2113.27 kHz
  Ideal LSF: 2108 kHz $-$ 369 kHz $=$ 1739 kHz
  Actual LSF: 2113.27 kHz $-$ 369.92 kHz $=$ 1743.35 kHz

For Master Group 2
  Ideal f_c: 6248 kHz
  Actual f_c: 4.01 kHz $\times$ 1562 $=$ 6263.62 kHz
  Ideal LSF: 6248 kHz $-$ 1739 kHz $=$ 4509 kHz
  Actual LSF: 6263.62 kHz $-$ 1743.35 kHz $=$ 4520.27 kHz

A 10-Hz variation caused the tone to be 11 kHz off at the radio channel output. Since each channel occupies a 4-kHz bandwidth, this drift offsets the tone approximately three channels in the composite baseband spectrum.

In Figure 8–6, super groups 25 through 28 include a prefix D in their super group number. This indicates that the carrier frequencies which these super groups modulate are *derived* and not produced by simply extracting harmonics of the 4-kHz base frequency. The carriers for super groups 15 through 18 are mixed with another lower harmonic (1040 kHz), and the upper side frequency is filtered for use as the carrier frequency for the D super groups. This is done to reduce the amount of phase jitter present in the higher super group carriers. Any phase jitter present in the 4-kHz base oscillator is multiplied, together with the frequency, when the carrier frequencies are generated. By deriving the higher super group carrier frequencies through a combination of multiplying and mixing, the total phase jitter is reduced. The phase jitter present in two signals that are mixed in a nonlinear device is not algebraically additive in the resultant. For example, if a 10-kHz signal with 5° of phase jitter is mixed with a 5-kHz signal with 5° of phase jitter, the upper frequency produced is 15 kHz, but it does not necessarily have 10° of phase jitter. The carrier frequencies for master groups and higher are also derived in a similar fashion.

Pilot Carrier

In the process of transmission, variations in signal amplitude may occur. The nature of these variations must first be determined before compensation can be made. A pilot frequency of 104.08 kHz is inserted at the group level. The derivation of this pilot frequency from the 4-kHz master oscillator signal is shown in Figure 8–8. Since five groups modulate five different pilot frequencies in the formation of super groups, each super group contains five different pilot frequencies (Figure 8–9).

As each group continues to be raised in frequency through modulation, the associated pilot frequency is also raised. At the receiver, it is assumed that each group experienced the same amplitude variations as its pilot frequency. Since the pilot is inserted at a known signal level, the variations that this signal sustained can readily be determined. As a result, these pilot frequencies are used for automatic gain control at all levels.

Figure 8–8 Derivation of the 104.08-kHz pilot frequency.

Figure 8–9 Pilot frequencies in (a) groups and (b) super groups.

Microwave Transmission

A radio channel comprised of three master groups (564 kHz to approximately 8.3 MHz), in order to be sent as a microwave (> 1 GHz) transmission, must still be raised further in frequency. An IF carrier frequency of 70 MHz is frequency modulated by the 564 kHz—8.3 MHz signal at a low modulation index (approximately 0.4). At such a low modulation index, frequency modulation produces only one pair of significant side frequencies for each frequency in the modulating signal. As a result, the frequency spectrum of the output resembles the frequency spectrum of an amplitude-modulated signal. This signal now amplitude-modulates a 6-GHz carrier frequency. The upper sideband that is produced is filtered and transmitted. This is shown in Figure 8–10. The bandwidth required to transmit 1800 VB channels is 16.6 MHz. Since FCC allows a 29-MHz bandwidth, the transmitted bandwidth is well within limits.

Baseband Signal

The composite FDM spectrum, whether it is a group, super group, master group, or jumbo group, is called the baseband signal. The *baseband signal* is the signal which modulates the signal that becomes the final carrier. In Figure 8–10, since the final carrier is 6.07 GHz, the frequencies 564 KHz to 8.3 MHz make up the baseband signal.

In FDM transmission, transmitted channels that leave a particular locale need not all be going to the same destination. Groups, super groups, master groups, and so on, may be broken down at various points, channels extracted, additional channels inserted, and the various groupings entirely reformed. If there are 1800 VB channels on a microwave circuit, only about 450 are anticipated to be used at a given instant: only 450 channels will have information going the same way at the same time. If there are 1800 ongoing conversations, only half (900) are expected to be speaking in the direction of transmission. When people talk, there are many pauses in their speech. Of the 900 people speaking in the direction of transmission, half (450) are expected to be actually talking at any given instant. Should all 1800 people actually speak in the same direction at the same time, the system would suffer a severe overload.

Figure 8–10   Raising frequencies to the gigahertz range.

FDM is used extensively by telephone companies, utilities, municipalities, the government, and the military to reduce the number of carrier facilities required in communication systems.

QUESTIONS

1. Define FDM.
2. The A-type channel bank frequency-division multiplexes incoming channels to form _____ .
3. Identify the number of voice band channels in each of the following:
   (a) Group   (b) Super group   (c) Master group
   (d) Radio channel   (e) Jumbo group
4. Identify the frequency range and the bandwidth at the output of each of the following:
   (a) Group   (b) Super group   (c) U600 master group
   (d) Radio channel made up of three U600 master groups
5. At what level is the 104.08-kHz pilot frequency inserted?
6. The 104.08-kHz pilot frequency is inserted for the purpose of supplying a carrier frequency to the receiver for demodulation purposes.   (T, F)
7. What is the difference between phase jitter and frequency drift?
8. The guard bands between super groups in a U600 master group all have the same bandwidth.   (T, F)
9. What is meant by a "radio channel"?
10. Amplitude modulation is used to generate FDM.   (T, F)
11. All carrier frequencies are generated solely as harmonics of the base frequency (4 kHz).   (T, F)
12. Why are the carrier frequencies for super groups D25, D26, D27 and D28 derived differently than the carriers for the other super groups?

PROBLEMS

1. Construct a table showing the lower sideband output frequency range for each of the following:
   (a) For the 12 channels in a group
   (b) For the 5 groups in a super group
   (c) For the super groups in a U600 master group
2. Location:
   A = group output

   B = super group output

   C = master group output

   D = radio channel output
   (a) What frequency range would a voice band channel occupy if it belonged to the following?

      (1) MG 1, SG 14, GP 2, Ch 11
      (2) MG 2, SG D27, GP, 4, Ch 3
  (b) Find the frequency of a 1-kHz tone belonging to MG 2, SG 18, GP 5, Ch 7.
  (c) Find the frequency of a 1.5-kHz tone belonging to MG 3, SG D26, GP 4, Ch 6.

		A	B	C	D
(a)	**(1)**	_____	_____	_____	_____
	(2)	_____	_____	_____	_____
(b)		_____	_____	_____	_____
(c)		_____	_____	_____	_____

3. What is the pilot carrier frequency at the master group output if it was derived from GP 1, SG 16, MG 2?

4. If a 4-kHz master oscillator drifted to 4000.1 Hz, what would be the frequency of a 1-kHz tone at the radio channel output for Ch 3, GP 4, SG 17, MG 2?

9

VOICE OR DATA TRANSMISSION WITH DIGITAL CARRIERS

DIGITAL COMMUNICATIONS

The boundaries encompassed by the phrase "digital communications" are not clear cut. *Digital communications* is commonly understood to be the transmittal of information from one point to another in digital form. Signals from voiceband channels are converted to digital pulses, transmitted on metallic lines (called *T-carriers* by the Bell system), and reconverted to analog signals at the receive end. The *source signal* can be *analog* (voice) or *digital* (data). In Chapter 7 it was shown that modems convert digital pulses into signals suitable for transmission on voice-grade lines. Therefore, digital data from a DTE can be converted to an analog signal by a modem, reconverted to digital (but of a different form), and then be transmitted as such on T-carriers. Digital information may be highly multiplexed. Multiplexed digital signals can PSK modulate an analog carrier and then be transmitted as a microwave signal. Microwave transmission of digital information is called *digital radio*. These various forms of digital transmission are depicted in Figure 9–1.

DDS

Additional digital services called DDS are provided by Telco. Unfortunately, these same initials represent two different types of services, Digital Data Service and Dataphone Digital Service. In *Digital Data Service,* the customer's digital signals are immediately placed on a T-carrier. This service was initially used to provide short-range digital communications. To extend the distance capability of this service, these digital signals are converted to seven-level PAM signals. The 1A-Radio Digital

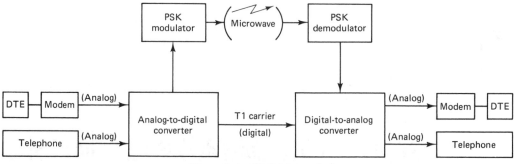

Figure 9–1 Digital transmission.

Terminal (1 A-RDT) then converts the multilevel PAM signals to a signal that occupies a bandwidth of approximately 500 kHz and transmits it as *data under voice (DUV)*. We have seen in Chapter 8 that FDM master groups are transmitted at frequencies of 564 kHz and above. A DUV signal is so called because it uses the same transmitting and receiving facilities as the master groups (Figure 9–2). With Digital Data Service, the signal is considered in digital form from the customer's originating equipment to the customer's terminating equipment. Again, allowances have to be provided in this terminology since neither the multilevel PAM nor the DUV signals are in digital form.

With *Dataphone Digital Service,* a modem is involved. Figure 9–3 contrasts Digital Data Service with Dataphone Digital Service.

Various DTEs, transmitting at 2400, 4800, 9600 bps, or 56 kbps, have modems that convert these digital data to analog signals which are then transmitted to a Telco central office. At the central office, modems reconvert these signals to digital, multiplex them with other dataphone circuits, and transmit the resulting signal at a 1.544-Mbps rate. Dataphone Digital Service differs from a "regular" digital channel in that the transmission rate is slower and voice signals are never multiplexed with the digital information.

Advantages of Digital Transmission

The major advantage of digital transmission is noise immunity. In the transmission of analog information, undesired amplitude, frequency, and phase variations, introduced

Frequency (kHz)

Figure 9–2 Data under voice (DUV).

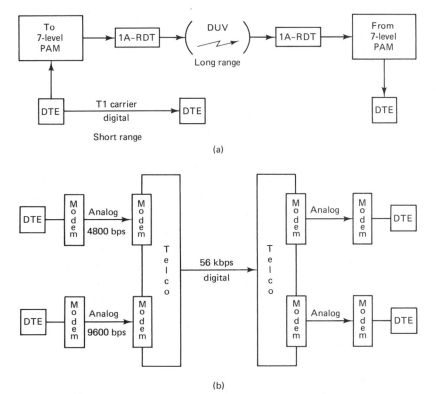

Figure 9-3   Digital data service and dataphone digital service: (a) digital data service; (b) dataphone digital service.

by either hardware or external noise sources, can produce errors in the received signal. With digital transmission, precise evaluation of the amplitude, frequency, and phase are not required. Although the digital pulses deteriorate as they travel down the transmission line, to extract the original information at the receive end, it is only necessary to determine whether the received signal at the time of sampling is either above or below a particular voltage threshold. *Regenerative repeaters* (not amplifiers), spaced approximately 6000 ft apart, detect the digital pattern, reconstruct the signal, and then retransmit it. The signal quality at the repeater output is equal to what it was when it was originally transmitted. The repeater distance of 6000 ft is chosen strictly from convenience. Digital pulses are transmitted along metallic facilities and these facilities must first be established. The use of existing facilities was the most convenient solution. Loading coils which are used to improve the quality of analog voice channel lines are spaced approximately 6000 ft apart. Telco manholes are available to access the lines at these points. Since these coils have to be removed to provide the BW required of T-carriers, it was logical to install repeaters in their place.

The trend toward digital communications is ever-increasing in the United States. Although not yet used for long-distance communications, there are more miles of T-carriers than miles of lines carrying analog signals. Previously, digital transmission was used for distances of 50 miles or less. Improved transmission techniques now permit digital transmission for distances of 500 miles with repeaters spaced approximately every 15,000 ft.

PULSE CODE MODULATION

Digital transmission is the transmission of voltages which represent logic 1's and 0's. There are several ways of transforming analog information into digital information:

1. *Pulse width modulation* (PWM) or *pulse duration modulation* (PDM):   The pulse width is proportional to the amplitude of the analog signal.
2. *Pulse position modulation* (PPM):   Varying the position of identical pulses with respect to a particular time reference.
3. *Pulse amplitude modulation* (PAM):   Converting analog amplitudes to discrete voltage levels. Although the resultant signal is made up of different discrete voltage levels, it cannot be considered digital information since these levels do not represent logic 1's and 0's. The output of the 2-to-4 level converter in the modulator circuit of Figure 7–29 is a PAM signal.

Pulse code modulation (PCM) is the most prevalent method of converting analog information to digital pulses. The analog voltage is sampled at regular intervals by a *sample-and-hold (S/H) circuit* and, while this sampled value of voltage is held constant, an analog-to-digital converter (ADC) converts it to a binary number. This number is proportional to the amplitude of the analog signal at the time of sampling. This is shown in Figure 9–4. The digital numbers will be explained shortly. The binary equivalent of the sampled voltage level is transmitted and reconverted at the receive end to a voltage level by a digital-to-analog converter (DAC).

Sampling Rate

The Nyquist sampling theorem establishes the minimum rate at which a signal must be sampled for accurate information reproduction. Essentially, if a signal is sampled at a rate (f_N) that is equal to or greater than twice the highest frequency (f_a) contained in the signal, the original signal can be accurately reproduced from these samples. If f_N is less than $2f_a$, distortion is introduced in the sample that will be carried through the system and appear in the demodulated output. This distortion is called *foldover distortion* or *aliasing*. Basically, a sampling circuit is a switch which is a nonlinear device. It has two inputs, the analog signal (f_a) and the rectangular sample pulse (f_N). A nonlinear device with two or more inputs acts as an AM modulator. f_N is amplitude-modulated by f_a. The two original frequencies (f_N and f_a), their sum ($f_N + f_a$), and their difference ($f_N - f_a$) are present at the output of the sample circuit. Since the sample pulse is rectangular and repetitive, it is made up of the fundamental frequency (f_N) and all of the higher harmonics of f_N. Therefore, f_N and all of its harmonics are amplitude-modulated by f_a. The output frequency spectrum is shown in Figure 9–5.

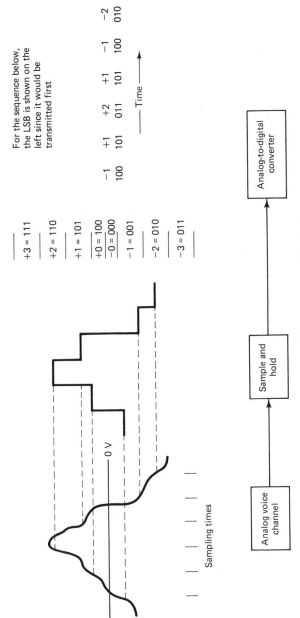

Figure 9-4 Pulse code modulation.

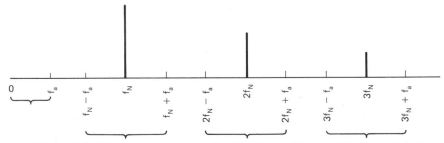

Figure 9–5 Frequency spectrum produced by the amplitude modulation of f_N and all of its harmonics by f_a.

Example

Maximum analog input frequency to a sampling circuit = 3 kHz. Sampling frequency = 6 kHz. The output will contain the original frequency range of 0 to 3 kHz and the band of frequencies 3 to 9 kHz produced by the amplitude modulation of the 6-kHz signal. These higher frequencies can be removed by filtering and the original signals will not be contaminated. If the sampling rate is reduced to 5 kHz, a frequency range from 2 to 8 kHz is produced by the modulation process. If frequencies above 3 kHz are filtered out, a band from 2 to 3 kHz, produced by modulation, still overlaps (folds over) the original frequency range. The presence of these frequencies in the original passband, after demodulation by the receiver, is foldover distortion. Since voice channels will be digitized for transmission, and a 0- to 4-kHz BW is allocated for each voice channel, a minimum sampling rate of 2×4 kHz = 8000 samples/second is required. A sample of the input signal must be taken every 1/800 s or every 125 μs. For the frequency range 0 to 4 kHz, since the highest frequency in this range (4 kHz) corresponds to the BW of the channel, the minimum sample rate rate = $2 \times$ BW (low-pass filter).

Sample and Hold

Aperture time is the time during which the analog input signal is being sampled. Since the analog signal may be changing during this time, if the ADC attempted to converted this varying voltage to digital, distortion would be introduced. This distortion is called *aperture distortion*. This distortion is eliminated by the use of sample-and-hold (S/H) circuits. In a sample-and-hold circuit, the aperture time is made very small. During this time, the analog voltage produces a charge across a capacitor. During the hold time, the capacitor voltage is converted into a binary number by the ADC. The sampled voltage is essentially nonvarying during the conversion process. In the S/H circuit of Figure 9–6, SW 1 is first closed and SW 2 is opened. SW 1 should provide a low-impedance output and a high current drive to allow the capacitor to adjust rapidly to the sampled analog voltage. At the end of the aperture time, SW 1 is opened and SW 2 is closed, coupling the capacitor voltage to the ADC. SW 2 should have a high input impedance to keep the capacitor voltage constant during the time it is being converted to a binary number by the ADC.

Quantization Noise

Between the maximum and minimum amplitudes of the analog signal, an infinite number of levels exist. In the encoding process, this infinite number of levels is

Figure 9–6 Sample and hold: (a) S/H pulse; (b) S/H circuit.

converted to a finite combination of binary numbers dependent on the number of bits allocated for each sample. Therefore, each binary number will have to represent a range of sampled voltages. At the receive end, each binary number is decoded to a single voltage level. The error introduced in the conversion process is called *quantization error* or *quantization noise*. As can be seen in Table 9–1 by comparing the maximum difference between the input and the decoded output voltages, the maximum error is equal to one-half of the voltage represented by the step value of the ADC (one-half of the voltage of the least significant bit (LSB)). Because of the high immunity to noise by a PCM signal, quantization noise is the major source of error in digital transmission.

If n bits are used, 2^n possible combinations are available at the output of the ADC. The most significant bit represents the sign: 1 for $+$, 0 for $-$. (In digital theory, the opposite scheme is used for signed binary number representation.) For $n = 3$, eight 3-bit combinations are available. Two of these combinations are used for zero: 100 for a positive zero and 000 for a negative zero. This leaves seven combinations (levels) for the DAC at the receive end to convert into a single voltage value. Since one of these bits is used as a sign bit, $n - 1$ bits remain to express the magnitude of the number. The maximum magnitude that can be obtained with n bits is $2^n - 1$. If $n = 3$ and 1 bit is used for the sign, 2 bits remain to represent the magnitude. The maximum number that can be expressed with 2 bits is $2^n - 1 = 3$. If the voltage of the LSB is 4 mV, the decoded voltage values are as shown in Table 9–1.

Since a range of input voltages is capable of producing each binary number, *the maximum value of this error is equal to one-half of the value of the LSB.* At the input, a range of 4 mV was assigned to each binary number. Note that although two combinations were used to express zero, the range of values expressing ± 0 is still 4 mV.

TABLE 9–1

Binary	Decimal	Decoded voltage value (mV)	Input voltage range (mV)
1 1 1	$+3$	12	10 to 14
1 1 0	$+2$	8	6 to 10
1 0 1	$+1$	4	2 to 6
1 0 0	$+0$	0	0 to 2
0 0 0	-0	0	-2 to 0
0 0 1	-1	-4	-6 to -2
0 1 0	-2	-8	-10 to -6
0 1 1	-3	-12	-14 to -10

Example

Given: 7-bit code (1 bit for sign, 6 bits for magnitude). LSB $= 0.15$ V.

(a) How many different voltage levels may be represented with this voltage scheme?
Ans.: $2^7 - 1 = 127$.

(b) What is the maximum positive output voltage that can be represented with this coding scheme?
Ans.: $1111111 = +63$. $63 \times 0.15 = +9.45$ V.

(c) What is the maximum peak-to-peak output voltage that can be represented with this coding scheme?
Ans.: $2 \times 9.45 = +18.9$ V.

(d) What is the voltage represented by the binary code of 1101010?
Ans.: $42 \times 0.15 = +6.3$ V.

(e) What is the voltage represented by the binary code of 0011010?
Ans.: $-(26 \times 0.15) = -3.9$ V.

(f) What is the binary code for $+4.75$ V?
Ans.: $4.75/0.15 = 31.7$. Round off to 32. The code is 1100000.

(g) What is the binary code for -2.72 V?
Ans.: $2.72/0.15 = 18.13$. Round off to 18. The code is 0010010.

Example

Given: Maximum peak-to-peak input voltage of 31.5 V. Six bits (5 for magnitude, 1 for sign) will be used for encoding.

(a) How many different voltage levels may be represented?
Ans.: $2^6 - 1 = 63$.

(b) Voltage of LSB?
Ans.: $31.5/63 = 0.5$ V.

(c) What is the maximum quantization error?
Ans.: 0.25 V; one-half of the LSB.

(d) What is the decoded value for 101101?
Ans.: $13 \times 0.5 = +6.5$ V.

(e) What is the decoded value for 011001?
Ans.: $-(25 \times 0.5) = -12.5$ V.

(f) To what binary number will $+13.62$ V be encoded?
Ans.: $13.62/0.5 = 27.24$. Round off to 27. The binary number is 111011.

(g) To what binary number will -9.37 V be encoded?
Ans.: $9.37/0.5 = 18.74$. Round off to 19. The number is 010011.

Resolution and Signal-to-Quantization Noise

The ADCs under discussion are linear—all of the voltage steps are equal. Therefore, the maximum amount of quantization error is the same for all input amplitudes.

However, all amplitudes do not have the same probability of occurring in normal speech. The majority of voice conversations have relatively low signal amplitudes. The *resolution* of an ADC or a digital-to-analog converter (DAC) identifies the smallest analog voltage value that can be discerned by these converters and is equal to the voltage of the least significant bit. For a given voltage range, the resolution can be increased by increasing the number of bits in these converters and decreasing the magnitude of the LSB. Ideally, nonlinear ADC would be preferred, with greater resolution provided at the lower amplitudes and less resolution at the higher amplitudes. The step output of such an ADC is shown in Figure 9–7.

For linear ADCs and DACs, the maximum quantization error is constant for all input amplitudes and is equal to one-half of the voltage of the least significant bit. The worst-case signal-to-quantization noise occurs when a minimum amplitude signal is applied—a signal whose voltage value is equivalent to the value of the least significant bit. In this case,

$$S/N \text{ (worst case)} = \frac{V(LSB)}{\text{quantization error}}$$

$$= \frac{V(LSB)}{V(LSB)/2} = 2$$

As the input signal amplitude increases, the signal-to-quantization error also increases. With nonlinear ADCs and DACs, the quantization error is increased for the larger analog input signal amplitudes. Therefore, the signal-to-quantization noise is more constant throughout the amplitude range of the input signal. Nonlinear ADCs and DACs do not improve the signal-to-quantization noise ratio for the lower-amplitude signals, but they do increase the accuracy (resolution) of these signals. This is accomplished at the expense of the signal-to-quantization noise ratio for the higher-amplitude signals. Unfortunately, nonlinear ADCs are not manufactured. Therefore, alternative approaches are required to achieve this result.

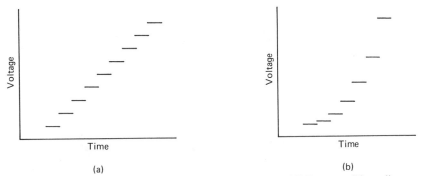

(a)   (b)

Figure 9–7   Linear versus nonlinear ADC: (a) linear ADC output; (b) nonlinear ADC output.

Analog Companding

Before conversion to digital, the analog signals are passed through a nonlinear amplifier which provides greater amplification for the lower-amplitude signals. This action compresses the analog voltage range from V_{max} to V_{min} and increases the probability of the presence of voltage amplitudes at all levels for the ADC. At the receive end, after the digital signal has been decoded, the amplitudes must be expanded—amplified nonlinearly, with the lower amplitudes amplified the least to restore the signals to their original proportions. This process of compression and expansion is called *companding*. Figure 9–8 shows a block diagram displaying analog companding. In the United States, μ-law is used for companding.

$$V_{out} = \frac{V_{max} \times \ln\left(1 + \mu V_{in}/V_{max}\right)}{\ln\left(1 + \mu\right)}$$

The amount of compression can be varied by varying μ. This is shown in Figure 9–9.

Example Using $\mu = 255$

For $V_{in} = V_{max}$, $V_{out} = V_{max}$, providing unity gain.
For $V_{in} = 0.75\,V_{max}$, $V_{out} = 0.948\,V_{max}$.
Gain $= V_{out}/V_{in} = 0.948\,V_{max}/0.75\,V_{max} = 1.26$.
For $V_{in} = 0.5\,V_{max}$, $V_{out} = 0.876\,V_{max}$, gain $= 1.75$.
For $V_{in} = 0.25\,V_{max}$, $V_{out} = 0.752\,V_{max}$, gain $= 3$.

From these examples, it can be seen that the lower amplitudes are amplified by a greater amount than the higher-amplitude input signals.

Dynamic Range

The *dynamic range* is a ratio of maximum to minimum expected voltage values in a system and is normally expressed as a dB value. It is an important factor in establishing the number of bits that will be used for PCM encoding.

Figure 9–8 PCM with analog companding.

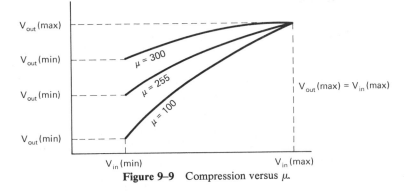

Figure 9–9 Compression versus μ.

Example

If a system has a dynamic range of 30 dB,

$$30 \text{ dB} = 20 \log \frac{V_{max}}{V_{min}}$$

$$\frac{V_{max}}{V_{min}} = 31.623$$

This is the number of conditions desired. To produce this number of conditions:

$$2^n = 31.623 \qquad \text{where } n \text{ is the number of bits}$$

$$n \log 2 = \log 31.623$$

Thus $n = 1.5/.3 = 5$ bits would be required.

V_{min} establishes the voltage of the least significant bit and the dynamic range affects the number of bits for each sample. There are other factors that must be considered which will further increase the number of bits per sample.

D1 Channel Bank

When digital transmission was first implemented, Telco's D (digital) channel banks performed the conversion of analog voice samples to digital pulses. The D1 channel bank used analog companding with a $\mu = 100$ and converted each voice sample into a 7-bit code (sign plus 6 bits for magnitude). An additional bit was added for signaling. *Signaling functions* include off-hook, on-hook, dialing, and so on. As increased requirements were placed on these digital transmission systems, improved resolution was required. One method of improving resolution is simply to increase the number of steps in the ADC. This would require that each sampled voltage be converted into a greater number of bits. If the rate of information transmission was to remain the same, the transmission bit rate would have to be increased. If a single sample is converted into 12 bits instead of 8, the transmission bit rate would have to be 1.5 times greater. If the transmission bit rate is increased, the time of each pulse is shortened, necessitating the use of a greater BW. This is not desired.

Digital Companding

D2, D3, and D4 channel banks use a $\mu = 255$ and convert each sample into 8 bits (sign plus 7 bits for magnitude). This is not totally true and the variations will be explained later.

Although D2, D3, and D4 channel banks use a $\mu = 255$, digital, not analog, companding is used to achieve this result. *Digital companding* involves compression at the transmit end after the signal has been sampled and expansion at the receive end before the received pulses are decoded. This is shown in the block diagram of Figure 9–10. Compare this figure with Figure 9–8 regarding the location where companding is accomplished.

Figure 9–10   PCM with digital companding.

In digital compression, each sample is first converted to a 12-bit binary number to obtain the greater resolution. However, instead of transmitting these 12 bits, they are first compressed to 8 bits and then transmitted. A compression of 12 bits to 8 bits cannot be accomplished without introducing error. Since increased resolution is desired for the lower amplitudes, this scheme minimizes the absolute error at these signal levels. In the 12-bit to 8-bit compression, the sign bit (MSB) is not affected. The remaining 11 bits are broken up into segments. Each segment is identified by a 3-bit binary number which is determined by subtracting the number of leading 0's of the 11-bit magnitude representation from 7.

Example

$$12\text{-bit code:}\quad \text{s}00001101010$$
$$\text{Segment code:}\quad 7 - 4 = 3 = 011$$

The digit following the leading 0's is understood to be a 1 (otherwise, there would be an additional leading 0) and is not transmitted. As seen in Table 9–2, segment zero is an exception. The next 4 bits following the first 1 are transmitted as is and make up the last 4 bits of the compressed digital number. Any remaining bits are truncated. This is shown in Figure 9–11.

When the 8-bit code is received, it must first be expanded back to a 12-bit code before decoding.

Figure 9–11   Twelve bit-to-eight bit code conversion.

Example

> Received 8-bit code = s 011 1010
>
> 7 − 3 = 4 leading 0's followed by a 1
>
> 12-bit code = s 0000 1 1010 xx

In this case, the receiver has no information concerning the two LSBs which could range from 00 to 11. To minimize this companding error, a leading 1 followed by 0's is inserted by the receiver to replace all truncated information so that the final 12-bit expanded code would be s00001101010. The absolute error due to companding will vary depending on the segment transmitted. Table 9–2 shows the original 12-bit values, the 8-bit compressed values, and finally the expanded values of these 8 bits.

TABLE 9–2   TWELVE BIT-TO-EIGHT BIT COMPRESSION

Segment	12-bit code	8-bit code	Expanded 12-bit code
0	s0000000abcd	s000abcd	s0000000abcd
1	s0000001abcd	s001abcd	s0000001abcd
2	s000001abcdx	s010abcd	s000001abcd1
3	s00001abcdxx	s011abcd	s00001abcd10
4	s0001abcdxxx	s100abcd	s0001abcd100
5	s001abcdxxxx	s101abcd	s001abcd1000
6	s01abcdxxxxx	s110abcd	s01abcd10000
7	s1abcdxxxxxx	s111abcd	s1abcd100000

a, b, c, and d are transmitted as is. Bits identified by x are lost in the companding process. Note that for segments 0 and 1, the original 12 bits are accurately reproduced, whereas for segment 7, only the 6 most significant bits have been recovered. Discounting the sign bit, the remaining 11 bits could produce $2^{11} = 2048$ combinations. Segments 0 and 1 can each be any of 16 different combinations, depending on the specific values of a, b, c, and d. In segment 2, the last 5 bits, a, b, c, d, and x, could range from 00000 to 11111, or 32 different combinations. However, in the companding process, these 32 combinations will result in only 16 levels, represented by a, b, c, d, and 1. Thirty-two levels have been compressed into 16. Similarly, for segment 7, 1024 levels have been compressed into 16. This nonlinear digital compression is shown graphically in Figure 9–12.

The 8-bit PCM code produced by digital compression has the longer sequence of consecutive zeros for the lower-amplitude signals. If the receiver encounters a long stream of zeros, it may lose bit synchronization, and the lower-amplitude voice signals are much more likely to occur. For this reason, the magnitude bits may be passed through an inverter prior to transmission. The sign bit remains unchanged. The

Segment +7 1024-to-16 level compression

Segment +6 512-to-16 level compression

Segment +5 256-to-16 level compression

Segment +4 128-to-16 level compression

Segment +3 64-to-16 level compression

Segment +2 32-to-16 level compression

Segment +1 no compression

Segment +0 no compression

Figure 9–12　Nonlinear digital compression (shown for positive values only).

resultant code is called *folded binary* and is shown in Figure 9–13. The obvious question now is: Why should long sequences of 0's cause the receiver to lose synchronization and not a long stream of 1's? This question will be answered in a later section where the types of pulses used in transmitting 1's and 0's are described.

The curve shown in Figure 9–12 closely approximates the μ-law curve for a $\mu = 255$. The amount of companding error introduced is proportional to the amount of compression. The smaller amplitudes are represented by smaller binary numbers and, for these numbers, there is very little if any compression. This was the goal of digital companding—to obtain the greater resolution offered by a 12-bit code for the smaller-amplitude signals. As the signal amplitude increases, the amount of companding error increases. However, the maximum amount of percentage error is constant for all segments.

```
1 111 1111              1 000 0000
    .                       .
    .                       .
1 000 0000              1 111 1111
           ──────▶
0 000 0000              0 111 1111
    .                       .
    .                       .
0 111 1111              0 000 0000   Figure 9–13  Folded binary.
```

The following formula is used for the computation of percentage error:

$$\% \text{ error} = \frac{\mid \text{Tx voltage level} - \text{Rx voltage level} \mid}{\text{receiver voltage level}} \times 100$$

The maximum percentage error will occur for the smallest number within a segment.

Example

For segment 3:

Transmit   s00001000000

Receive   s00001000010

$$\% \text{ error} = \frac{\mid 64 - 66 \mid}{66} \times 100 = 3.03\%$$

For segment 7:

Transmit   s10000000000

Receive   s10000100000

$$\% \text{ error} = \frac{\mid 1024 - 1056 \mid}{1056} \times 100 = 3.03\%$$

Time-Division Multiplexing

In TDM, the information from the various channels are interleaved and separated by time. This is shown in Figure 9–14. The various channels that are to be multiplexed are alternately sampled and transmitted. PCM easily lends itself to TDM. Telco's D channel banks accomplished the sampling (Figure 9–15), encoding, and multiplexing. Twenty-four voiceband channels are each sampled at a 8000-sample/second rate (at

Figure 9–14   TDM.

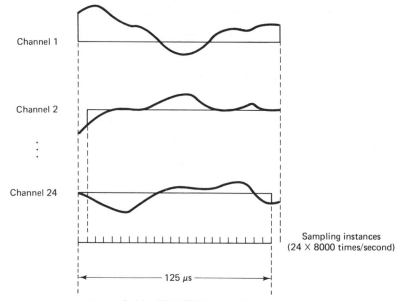

Figure 9–15 PCM/TDM sampling sequence.

the same rate but not at the same time) and each sample is encoded into an 8-bit number.

$$24 \text{ samples} \times 8 \text{ bits/sample} = 192 \text{ bits}$$

For every 192 bits, an additional bit, the *frame bit*, is added for the purpose of synchronization. The resulting 193 bits make up a *frame* (Figure 9–16). Since frames are generated at the same frequency as the sampling rate, time-division multiplexing 24 channels requires a transmission rate of $193 \times 8000 = 1.544$ Mbps.

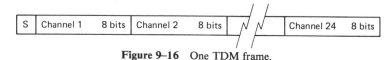

Figure 9–16 One TDM frame.

The D1 channel bank used analog companding with a $\mu = 100$ and a 7-bit code for the sample. A signaling bit was transmitted with every 7-bit code for every channel. The frame bits transmitted by the D1 channel bank simply alternated between 1 and 0. Signaling bits are not required that often. With the D2, D3, and D4 channel banks, digital companding is used with a $\mu = 255$ and 8 bits are used for each sample. Frames are grouped in units of 12. In the sixth and twelfth frames, the least significant PCM bit of every channel is used as a signaling bit, leaving only 7 bits for data information in these frames. The framing bits now must not only provide the necessary synchronization information but also must identify the sixth and twelfth frames. This is accomplished by using the odd-channel frame bits for synchronization and the even-channel frame bits to identify the sixth and twelfth frames. These frame bit patterns are shown in Figure 9–17. If only the framing bits are observed, the sixth frame (A-channel signaling) is identified

Frame number	1	2	3	4	5	6	7	8	9	10	11	12
Synch bits	1		0		1		0		1		0	

6th and 12th frame identification 0   0 ———▶ 1   1   1 ———▶ 0

Figure 9–17 Frame bit sequence.

by a 0-to-1 transition and the twelfth frame (B-channel signaling) is identified by a 1-to-0 transition.

The outputs from a D channel back are placed on T-carriers. In order to be able to maintain timing, the D-channel banks are required to ensure that no sequence is transmitted that has more than 14 consecutive zeros. Each encoded sample (8 bits) is monitored for the presence of at least one nonzero bit. If such a bit is not present, a 1 is substituted for the bit in bit position 7 (bit positions numbered 1 through 8).

Worst-Case Examples

  (a) 1000 0000   0000 0001

      No substitution, 14 consecutive 0's, a 1 present in encoded sample.

  (b) 1000 0000   0000 0000

      A 1 would be substituted for bit 7 of the second encoded sample:

      1000 0000   0000 0010 (order of arrow is to the left)

Digital Channel Banks

The D1 channel bank sequentially samples the analog signal of 24 voiceband channels. Common equipment is used to perform analog compression ($\mu = 100$) on each sample. The resultant signal is then converted into a 7-bit code (sign plus 6 bits for magnitude). An additional bit, the signaling bit, is added to every 7-bit code of every channel. These channel banks are now obsolete. Saying that they are obsolete does not necessarily imply that they no longer exist. When opportunity permits, Telco is replacing them with D3 channel banks.

D2 and D3 channel banks are functionally similar. However, the D3 channel banks are the first to incorporate integrated circuitry which reduces the size and the power consumption required. Both D2 and D3 channel banks time-division multiplex 24 channels and use digital companding which approximates the $\mu = 255$ curve. Eight bits (sign plus 7 bits for magnitude) are used to represent each sample, with the least significant bit stolen every sixth frame for signaling.

The D4 channel bank uses customized LSI circuitry and time-division multiplexes 48 voice channels using digital companding with $\mu = 255$. Rather than use common equipment for encoding, each channel has its own encoder/decoder. The D4 channel bank multiplexes the encoded signals. The obvious advantage is that, should an encoder malfunction, only one channel is lost. The recent decreases in cost of *codecs* (coder/decoder) has made individual channel encoding/decoding cost-effective. The ultimate goal is to digitize voice transmission entirely. Rather than have an analog signal transmitted from the telephone to the digital channel bank, by incorporating a codec right into the telephone set, the voice signal is digitized at the outset. The transmission rate of the D4 channel bank is 3.152 Mbps. This bit rate is slightly higher than twice the output bit rate of a 24-channel multiplexed output. Since a larger number of channels are multiplexed, additional control bits are required.

North American Digital Hierarchy

Although the transmission bit rate at the output of the D1, D2, D3, D4 channels banks is either 1.544 Mbps or 3.152 Mbps, these outputs are further multiplexed to achieve an even greater transmission bit rate. Table 9–3 shows the line nomenclature and the type of signal transmitted on that line.

Example

> A DS-1 signal has a transmission rate of 1.544 Mbps. This signal represents samples from 24 voice band channels time-division-multiplexed together. The DS-1 signal is transmitted on a T1 line.

Multiplexers and Cross-Connects

Figure 9–18 shows the multiplexers used to achieve this higher transmission rate. Bell's nomenclature is used to identify the multiplexers. These multiplexers are sometimes called *muldems* (multiplexer/demultiplexer) since they must handle signal conversion in either direction. For simplicity, the number of inputs shown to each multiplexer reflects the number required from the multiplexer of the next lower order. This is not a restriction—M13 may receive either DS-1 or DS-2 or the proper combination of both to provide a DS-3 output.

DS-1, DS-2, DS-3, and DS-4 interface points are called the DSX-1, DSX-2, DSX-3, and DSX-4 cross-connects, respectively. The Bell system provides detailed specification for data stream requirements at these cross-connect points, namely, how the data stream appears at the end of a definite length of a particular cable. Maximum distances to cross-connect points are shown in Figure 9–19. These specifications set the maximum separation distance between the digital equipment. This is shown in Figure 9–20.

TABLE 9–3   NORTH AMERICAN DIGITAL HIERARCHY

Line type	DS (digital signal)	Bit rate (Mbps)	Number of VB channels
T1	DS-1	1.544	24
T1C	DS-1C	3.152	48
T2	DS-2	6.312	96
T3	DS-3	44.736	672
T4M	DS-4	274.176	4032

Figure 9–18 North American Digital hierarchy.

Cross-Connect	Maximum Distance (ft)
DSX-1	750
DSX-2	1000
DSX-3	450
DSX-4	150

Figure 9–19   Maximum distances to cross-connects.

Digital Line Signals

When information is transmitted in digital form, various factors must be considered:

1. Timing recovery
2. Bandwidth required
3. Ease of detection
4. Error detection

Some of the choices for pulse format are shown in Figure 9–21.

Transmission voltages.   Unipolar transmission involves the transmission of only a single nonzero voltage level ($+$ V for a 1, 0 V for a 0). If the bit voltage is maintained for the entire bit time, the waveform is termed *Non-Return-to-Zero* (NRZ). In *Return-to-Zero* (RZ) transmission, the bit voltage is maintained for only a portion of the bit time. RZ transmission has an associated *duty cycle* which identifies the

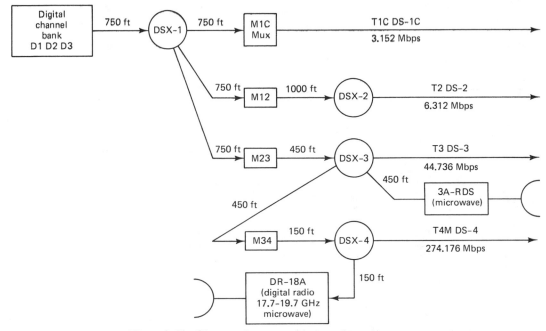

Figure 9–20   Distances between equipment dictated by cross-connects.

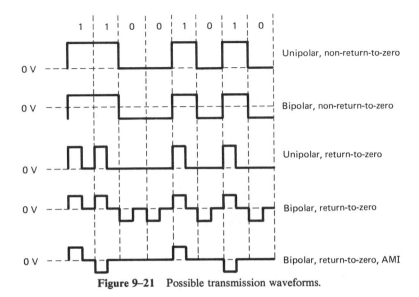

Figure 9–21 Possible transmission waveforms.

percentage of bit time that a nonzero polarity is maintained. Since a 50% duty cycle is used with RZ transmission, this duty cycle will be assumed in all future references to RZ. In *bipolar transmission,* two nonzero voltage levels are involved—a positive value for a 1 and a negative value for a 0. Bipolar transmission may be NRZ or RZ. Because of clear-cut advantages, T1 and T1C carriers use bipolar RZ with *alternate mark inversion* (AMI). In this line format, 0's are transmitted at a 0-V level and 1's alternate in polarity.

Timing recovery.   For timing recovery, a sufficient amount of transitions in the transmitted signal is essential. If NRZ transmission is used without any modifications, long strings of 1's and 0's would produce a constant line voltage. Since these would make timing recovery extremely difficult, unmodified NRZ is not a suitable form for digital transmission. RZ transmission forces transitions in a constant string of 1's. These transitions are absent in RZ for a constant string of 0's if a unipolar format is used.

Bandwidth requirements.   Line signals undergo deterioration en route to a repeater or to the final destination. To reconstruct the digital signal, the received signals are filtered to extract the fundamental frequency and then sampled. The low-pass filters pass the highest fundamental frequency of the incoming signal. The fundamental frequency of any rectangular waveform is the reciprocal of the time for one cycle no matter what the duty cycle (Figure 9–22).

Digital pulses are described as having a *(sin x)/x frequency spectrum.* If the factors of this ratio are considered separately, its meaning is much more readily understood. The numerator is a sine wave whose instantaneous amplitudes depend on *x;* the denominator increases with *x.* Therefore, this ratio simply represents a dampened sine wave. A square wave is made up of the fundamental frequency and an

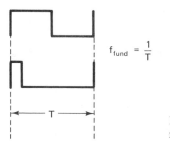

$f_{fund} = \dfrac{1}{T}$

Figure 9–22   Fundamental frequency of rectangular waveforms.

infinite number of odd harmonics whose amplitudes decrease with frequency. If these frequencies are plotted (Figure 9–23), their envelope would have a $(\sin x)/x$ curve. A dc component would be present if the square wave is not symmetric around the 0-V level. By filtering out the fundamental frequency and then sampling it at the time of maximum amplitudes, the binary 0–1 values can be extracted.

In NRZ transmission, the fastest line variations occur when alternate 1's and 0's are transmitted. From Figure 9–24 it can be seen that RZ require a bandwidth equal to the transmission bit rate. RZ introduces variations in the line signal which NRZ does not have, but it requires twice the BW. Bipolar RZ with AMI combines the best of both worlds. The fastest variations occur when consecutive 1's are transmitted. Since these 1's will be of opposite polarity, the BW, as seen in Figure 9–25, is one-half of the bit rate, and line variations are introduced when consecutive 1's are transmitted.

Ease of detection.   Unipolar transmission involves the transmission of only a single nonzero voltage level. To extract the digital data from the received signal, decision circuitry is required to determine whether a signal value is either above or below a positive threshold value. A long string of 1's (NRZ or RZ) may produce variations in the threshold value which may cause erroneous conversions. Variations in the threshold values may also be encountered in the bipolar transmission, again caused by long strings of 1's and 0's. With RZ AMI transmission, since all nonzero line voltages are produced by 1's and since these 1's alternate in polarity, the average dc line voltage is 0 V. Therefore, for this type of transmission, line signals have very little effect on the decision threshold levels.

Error detection.   By the very nature of RZ AMI transmission, received pulses should alternate in polarity. If two successive pulses of the same polarity are encountered, a noise impairment is definitely present: either a pulse was sufficiently

$\dfrac{\sin x}{x}$

Figure 9–23 $(\sin x)/x$ frequency spectrum.

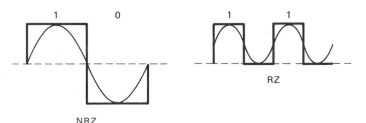

Figure 9–24   Highest fundamental frequency for RZ and NRZ.

reduced in amplitude to eliminate a 1 or noise in a 0-V region was of sufficient amplitude to be interpreted as a 1.

T2 and T3 carriers essentially use the same form of line signals as those used by the T1 and the T1C carriers. Bipolar RZ AMI transmission creates variations in long strings of consecutive 1's. However, long strings of 0's could still present a problem in trying to maintain timing. Earlier in this chapter it was seen that in a multiframe transmission, provisions are taken to prevent more than 14 consecutive 0's from being transmitted. The burden of not transmitting more than 14 consecutive 0's is placed on the source (D channel banks). Since the DS-2 signal on a T2 carrier is constructed by multiplexing four DS-1 signals and the DS-3 signal on the T3 carrier is comprised of multiplexed DS-1 and DS-2 signals, alternative measures are taken to eliminate long zero strings on these lines.

B6ZS stands for *Binary Six Zero Substitute.* Whenever six 0's are encountered, a substitution of either $0 - + 0 + -$ or $0 + - 0 - +$ is made for them. Here $+$ and $-$ represent the polarity of the voltage representing a 1. These substitutions purposely cause the transmission of consecutive 1's with the same polarity. If the pattern produced can be recognized by the receiver, the receiver can replace the substituted pattern with 0's. The substitute pattern chosen should produce a pulse of the same polarity at the second and fifth bits. This then would depend on the polarity of the last 1 preceding the string of six 0's.

Example

$+$ and $-$ are used to indicate the polarity of the 1 bit.

(a) Original: $+ 000 - 0 + 0000 - 0000000 + \cdots$
    With substitution: $+ 000 - 0 + 0000 - 0 - + 0 + - 0 + \cdots$
(b) Original: $+ 00000 - 00 + 0000000 - 00 + \cdots$
    With substitution: $+ 00000 - 00 + 0 + - 0 - + 0 - 00 + \cdots$

B3ZS stand for *Binary Three Zero Substitute.* If DS-2 signals are multiplexed

Figure 9–25   Fundamental frequency of bipolar RZ with AMI.

with DS-1 signals to form a DS-3 signal, the B6ZS must first be detected and stripped from the DS-2 signal and, before transmission on a T3 carrier, substitutions are made for every three consecutive 0's. The substitute patterns are one of the following: 00 +, 00 −, − 0 −, or + 0 +. The pattern chosen should cause a polarity repetition at the third substitute bit. The pattern chosen is dependent not only on the polarity of the 1 immediately preceding the three consecutive 0's but also on whether the number of 1's transmitted since the last substitution is odd or even. 00 − and 00 + are substituted if an odd number of 1's were transmitted since the last substitution.

Example

Original Transmission:

$$0 + 0000 - 0 + 0000 - 00 + - 0000 + 00 - + - 00000 + \cdots$$

With B3ZS:

$$0\ 1\ \underline{000}\ 0\ 1\ 0\ 1\ \underline{000}\ 0\ 1\ 00\ 1\ 1\ \underline{000}\ 0\ 1\ 00\ 1\ 1\ 1\ \underline{000}\ 00\ 1\ ...$$
$$0 + \underline{00+}\ 0 - 0 + \underline{-0-}\ 0 + 00 - + \underline{00+}\ 0 - 00 + - + \underline{-0-}\ 00 +$$
$$1 \qquad 1\ 2 \qquad 1 \quad 2\ 3 \qquad 1 \quad 2\ 3\ 4$$

As mentioned earlier, unmodified NRZ is generally unsuitable for binary transmission. On T4 carriers, NRZ transmission is used but the data are first scrambled. Scrambling was discussed in Chapter 7. Figure 9–26 lists the line signals of the various T-carriers.

T1, T1C	Bipolar RZ AMI
T2	B6ZS
T3	B3ZS
T4	Scrambled unipolar NRZ

Figure 9–26   Line signals used by various T-carriers.

DELTA MODULATION

Delta modulation is an alternative method of digitizing analog information. Delta modulation is related to digital encoding in a manner similar to the way DPSK is related to PSK. Rather than determine the binary code value solely from the sampled value of voltage, the binary output is dependent on whether the present sample is more or less positive than a reference signal which is determined from a history of previous sample sizes. If the present sample is more positive than this reference, a logic 1 is outputted; if the present sample is less positive than the reference, a logic 0 is outputted. The major advantage of delta modulation is the simplicity in encoding and decoding and the associated cheaper cost. However, the encoded signals cannot be integrated with other digital signals which use the D channel banks, nor do these signals lend themselves easily to being multiplexed. A block diagram of a linear delta modulator is shown in Figure 9–27.

As long as the analog input remains more positive than the reference voltage supplied by the digital-to-analog converter, the Up Enable signal for the counter will be active and the DAC output will increase in step increments. The rate at which the steps

Figure 9–27   Linear delta modulator.

change values (the width of a step) is determined by the clock frequency; the size of one step is determined by the resolution of the DAC and its input voltages. In the idealized situation shown in Figure 9–28, the DAC output follows the input voltage. At the

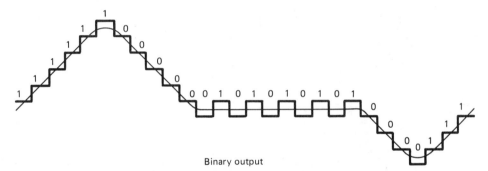

Binary output

Figure 9–28   Ideal action by a linear delta modulator.

receive end (Figure 9–29), the received 1's and 0's can be used to reproduce the staircase waveforms and the low-pass filter would smooth out this waveform recreating the analog waveform.

Figure 9–29   Delta demodulator.

Two problems arise with delta modulation—quantization or granular noise and slope overload—and each requires contrasting cures. *Granular noise* is the difference between the value of the sampled voltage and the value of the step size. In Figure 9–30, if waveform A is the original voltage and waveform B is the waveform reconstructed from the received digital data, the reproduced waveform possesses greater variations

B, reconstructed signal

A, original signal

Figure 9–30   Distortion due to granular noise.

than were present in the original signal. This problem can be alleviated by increasing the resolution of the ADC (making the step sizes smaller) and increasing the clock frequency.

Granular noise is more prevalent in analog signals whose amplitudes vary slowly. *Slope overload* occurs when the analog voltage varies at a rate faster than the delta modulator can maintain. This is shown in Figure 9–31. This problem would be solved if

Figure 9–31   Slope overload.

Figure 9–32   Continuous slope delta modulator.

the step size was increased so that the staircase waveform could catch up faster with the analog signal. To minimize granular noise, a small step size is desired; to minimize problems presented by slope overload, a large step size is desired. Obviously, for linear delta modulators, a compromise in step size must be achieved.

Adaptive delta modulators (ADM) automatically vary the step size based on signal requirements. *Continuous variable-slope delta modulator (CVSD)* is one form of an adaptive delta modulator. A block diagram of a CVSD is shown in Figure 9–32. Whenever an output sequence of either four 1's or four 0's is detected (an indication that the ADC output is trying to catch up with either a continuously rising or falling voltage), the step size is automatically increased. When this condition is no longer present—indicated by four nonsimilar bits in the shift register—the step size is returned to normal. Various other algorithms are available for adaptive delta modulators.

QUESTIONS

1. What is the difference between Digital Data Service and Dataphone Digital Service?
2. A frequency range of 3 to 10 kHz is to be sampled for transmission. What is the minimum sampling rate required for the original signal to be reproduced at the receive end?
3. The voltage in a S/H circuit is *continuously* adjusting to the analog input voltage.   (T, F)
4. The analog sample is converted to a digital number during the aperture time.   (T, F)
5. What is the cause of quantization noise?
6. What is the major advantage of digital communications?
7. If the resolution of an ADC is increased, (*more, the same amount, fewer*) bits would be used for each sample.
8. What is meant by a linear ADC?
9. In μ-law compression, the maximum input voltage appears (*amplified, unaffected, attenuated*) at the output.
10. In μ-law compression, a voltage less than the maximum appears (*amplified, unaffected, attenuated*) at the output.
11. In μ-law compression, the gain increases as the amplitude of the input signal _____ .
12. In μ-law compression of an analog signal, what is compressed?
13. All D channel banks use analog compression.   (T, F)
14. Since the formula for μ-law compression involves analog values, why is digital compression ascribed the name $\mu = 255$?
15. Explain fully how a DS-1 signal (1.544 Mbps) is achieved.
16. All D channel banks output signals from 24 multiplexed voice band channels.   (T, F)
17. A muldem is an acronym for _____ .
18. As the line bit rate increases, the number of consecutive zeros allowed decreases.   (T, F)
19. What does AMI stand for?

Delta Modulation

20. Slope overload occurs when the staircase voltage of the modulator is changing faster than the analog signal.   (T, F)

21. Granular noise is the same as quantization noise.   (T, F)
22. Increasing the step size would increase:
    (a) The probability of slope overload.   (T, F)
    (b) Granular noise.   (T, F)
23. Digital pulses outputted by the delta modulator vary in amplitude proportional to the difference in the previous signal amplitude samples.   (T, F)
24. What determines the sampling rate in a delta modulator?
25. What does "CVSD" stand for?
26. CVSD is a form of adaptive delta modulation.   (T, F)
27. In the CVSD discussed, if five consecutive 1's are transmitted, the modulator has been trying to reach the analog voltage by increasing the staircase voltage in five equal steps.   (T, F)

PROBLEMS

1. Analog compression is used with $\mu = 255$.
   (a) If $V_{max} = 4.032$ V, what is the corresponding compressed voltage value?
   (b) If $V_{min} = 2$ mV, what is the corresponding compressed voltage value?
   (c) If 6 bits are used to encode V_{max}, what is the voltage value of the LSB?
   (d) If a computer is available,
       (1) Manipulate the μ-law formula so that given V_{out}, V_{in} may be computed.
       (2) Write a program to print out V_{in}, V_{out} and gain for V_{out} values ranging from 0.064 to 4.032 V in increments of 0.064 V.

   The following questions may be answered from the computer printout or they may be calculated individually.

   (e) What was the gain if the compression output was 0.064 V?
   (f) What was the gain if the compression output was 4.032 V?
   (g) Is the amplification linear?
   (h) What is the uncompressed value for $V_{max} - V_{min}$?
   (i) What is the compressed value for $V_{max} - V_{min}$?
   (j) For V_{out} from 0.192 to 0.32 V, the difference is 0.128 V. What is the voltage difference for the two corresponding V_{in} values?
   (k) What is the difference in the two V_{in} values corresponding to V_{out} from 3.968 to 4.032 V?
   (l) Are smaller or larger amplitudes compressed more?
2. If PCM used 6 bits for the code (1 bit for the sign, 5 bits for magnitude) and the LSB = 0.2 V:
   (a) How many different voltage levels could be represented by this scheme?
   (b) What is the maximum positive output voltage that can be obtained with this scheme?
   (c) What is the voltage represented by 1 0 0 1 1 1 ?
   (d) What voltage is represented by 0 1 0 1 0 1 ?
   (e) What is the code for +0.88 V?
   (f) What is the code for −2.15 V?
3. If the maximum input voltage to a PCM system was 12.7 V and 7 bits (1 bit for the sign, 6 bits for magnitude) were used for encoding:

(a) What is the voltage of the LSB?

(b) What is the maximum possible error voltage?

(c) What is the total number of output voltages that can be obtained?

(d) What is the voltage represented by 1 0 0 1 1 0 1 ?

(e) What is the voltage represented by 0 1 1 0 1 1 1 ?

(f) What is the code for + 4.2 V?

(g) What is the code for − 3.6 V?

4. PCM/TDM system multiplexes 24 voice band channels. Each sample is encoded into 7 bits and 1 bit is added to each frame as a framing bit. The sampling rate is 9000 samples/second.

(a) What is the time of one frame?

(b) What is the time of one bit?

(c) What is the bit transmission rate?

5. PCM/TDM system multiplexes 32 voice band (0 to 4 kHz) channels. Each sample is encoded into 8 bits. There is one frame bit.

(a) What is the time of one frame?

(b) What is the time of one bit?

(c) What is the bit transmission rate?

6. A 12-bit code has been digitally compressed into 8 bits. The LSB = 0.03 V. For an analog signal of 1.74 V:

(a) What is the 12-bit code?

(b) What is the compressed 8-bit code?

(c) What is the value of the received decoded voltage?

(d) What is the percentage error in the transmission?

7. If the received compressed 8-bit code is 1 1 0 1 1 1 0 1, to what 12-bit code is it converted?

8. LSB = 0.02 V for a 12-digit linear code. What is the input voltage range that would be converted to:

(a) 1 0 0 0 0 0 0 0 0 0 0 1 ?

(b) 0 0 0 0 0 0 0 0 0 0 0 1 ?

(c) 1 0 0 0 0 0 0 0 0 0 0 0 ?

(d) 0 0 0 0 0 0 0 0 0 0 0 0 ?

(e) 1 0 0 0 0 1 0 1 0 0 0 1 ?

(f) 1 0 0 0 0 1 0 1 0 0 1 0 ?

9. For each of the following 12-bit codes, find the input voltage range and the corresponding digitally compressed 8-bit code. LSB = 0.02 V.

(a) 1 0 0 0 0 0 0 0 1 0 0 0

(b) 1 0 0 0 0 0 0 0 1 0 0 1

(c) 1 0 0 0 0 0 0 0 1 0 1 0

(d) 1 0 0 0 0 0 0 0 1 0 1 1

(e) 1 0 0 0 0 0 0 0 1 1 0 0

(f) 1 0 0 0 0 0 0 0 1 1 0 1

(g) 1 0 0 0 0 0 0 0 1 1 1 0

(h) 1 0 0 0 0 0 0 0 1 1 1 1

(i) What is the maximum quantization error for each of these?

(j) When the receiver converted the received 8-bit codes, expanded them back to 12 bits, to what voltage values would these 12 bits be converted?

(k) In the transmission of the codes above, no error would be introduced as a result of companding.   (T, F)

10. Assuming that the previously transmitted 1 had a positive polarity, draw the RZ, AMI waveform for the following bit sequence:

$$1\ 1\ 0\ 0\ 1\ 1\ 0\ 0$$

11. The following bit sequence is to be transmitted on a T2 carrier. Make the necessary substitutions to prevent excessive strings of 0's.

$$+\ 00000000\ -\ 0000000\ +\ 00000\ -\ 0000000\ +$$

12. The following bit sequence is to be transmitted on a T3 carrier. Make the necessary substitutions to prevent excessive strings of 0's.

$$100001000010011000001000010010000100001000$$

10

ASYNCHRONOUS PROTOCOL

Asynchronous transmission is characterized by the use of start and stop bits to frame each transmitted character (Chapter 5). For the typical system described, ASCII will be used as the character code. Asynchronous modems such as the 202S and the 202T are used to achieve half- and full-duplex operation (Chapter 7). As stated in Chapter 2, line protocol is a set of rules governing the transmission of information. The line protocol described will be that used in a multipoint system.

The primary function of the *line control unit* (LCU) is to control the flow of all transmission. No remote station may transmit unless it is first allowed to do so by the LCU. A *poll* is an interrogation of the remote station by the LCU to see if it has any information to send. A *selection* is an interrogation of the remote station by the LCU to see if it is ready to receive a message from the primary. Each remote station has its own unique *Transmitter Start Code* (TSC). When a station receives its own TSC, it is being polled. A TSC may consist of two ASCII characters, such as DC3 A. The first control character simply indicates that what follows is a unique remote TSC. Other systems may require that each two-character ASCII TSC be followed by DEL as a TSC identifier. The former symbology will be used in the examples in this text. Although DC3 and DEL stand for Device Control 3 and DELETE in the ASCII tables, remember that these are only 7-bit codes and any device or system may use these codes to represent or do whatever the system designer wishes. Each remote station has one or more *Call Directing Codes* (CDCs). These are not necessarily unique. When a remote station receives one of its CDCs, it is being asked by the LCU if it is ready to receive information—it is being selected. If several remotes had the same CDC, the LCU could select and send the same message to all of them simultaneously. If more than one station but not all possess a common CDC, that CDC is called a *group address*. If all stations have a common CDC, that CDC is called a *broadcast address*. Obviously, TSCs, by their nature, must be unique since the LCU would produce chaos if it could poll several remotes simultaneously and they all had something to send.

Remote Terminal Equipment

The following equipment may be found at a remote terminal: a display unit (CRT), keyboard, and/or a read-only printer (ROP). A *station controller* (Staco) serves as an interface between this equipment and the communications link.

Display (D). A CRT is normally capable of displaying 24 lines with 80 characters/line. This implies that there must be $24 \times 80 = 1920$ bytes of memory associated with the CRT to store the ASCII code for each character shown. Additional display memory may provide storage for three such pages (72 lines). The first line of each page is preceded by one, two, or three dots to identify the page viewed. The information displayed on the CRT can be highlighted and/or protected. If a character space is *protected,* the *keyboard operator* cannot, in normal operation, overwrite that character.

Keyboard (K). Figure 10–1 shows a typical asynchronous keyboard. The keyboard may be a stand-alone device or it may be connected with the display (KD) and/or printer (KDP). When in KD configuration, whatever is typed is displayed on the CRT. The keys at the left are for cursor control on the CRT. The HOME position on the CRT is the first row, first column. Note that the uppercase characters for the letters A through Z are data link control characters. These characters may be obtained by depressing CONTROL and then hitting the appropriate key. The terminal may be in one of three modes, SEND, RECEIVE, or LOCAL (upper left on keyboard). If the terminal is on-line with no information to send, it would normally be in the REC mode. The Staco monitors the terminal mode and responds accordingly when polled or selected by the LCU. If the station is polled while the terminal is in the REC mode, the Staco would respond with: "I have no message to send. I am ready to receive." In response to a selection, the Staco would respond with: "I am ready to receive." The terminal is placed in the LOCAL mode if the operator is composing a message on the CRT. At this time, the typed information is simply entered into display memory and displayed on the CRT, but it is not transmitted. If the Staco is polled or selected while the terminal is in LOCAL, the Staco would respond with: "I have no message to send. I am not ready to receive." The LCU must receive positive acknowledgment before transmitting any information to the remote terminal. In the case just cited, if the LCU transmitted information while the terminal was in LOCAL, the transmitted information would be displayed on the CRT, destroying the message that the terminal operator was trying to compose. After the operator completed typing a message, he or she would enter the SEND mode. The message itself would not be transmitted until that station was polled by the LCU. If the station was selected while the terminal was in the SEND mode, the Staco would respond with: "I am not ready to receive. I have a message to send."

Figure 10–1 Typical asynchronous keyboard.

(KDP only)

| SEND | REC | LOCAL | | | FORM SEND | | PRINT ON LINE | PRINT LOCAL | | | | | HIGH LIGHT | FORM ENTER | TAB SET | TAB CLEAR | | CLEAR |

(Full ASCII only)

Depressing the PRINT LOCAL causes all information in display memory from the current cursor location to the control character that would halt transmission to be printed on the printer. If the PRINT ON LINE key is depressed while the terminal is in the REC mode, all received information is simultaneously printed and displayed on the CRT. If the PRINT ON LINE key is depressed before the operator enters the SEND mode, information in display memory will be printed as it is being transmitted.

The FORM ENTER key is used in the preparation of a form that will be used repeatedly. After the form is completed, depressing the FORM ENTER key a second time will get the operator out of the FORM ENTER mode. The form characters are protected and nonhighlighted. If the operator now fills out the form, the data used for form completion appear highlighted on the screen and are not protected. There are many KDs and each KD has various options. A KD may be optioned to transmit only the unprotected data in display memory. With this option, only the information entered to complete the form would be transmitted. This assumes that the primary already has a copy of the form and needs only the entered data. By depressing the FORM SEND key and then the SEND key, both the form and the entered data (protected and unprotected information) would be transmitted. The terminal operator may also want the primary to store the form for later retrieval. In this case, the operator would depress the FORM SEND key and then the SEND key with just the form displayed on his CRT.

Figure 10–2 shows a block diagram of the terminal equipment. Figure 10–3 shows the signal connections of the Staco to the terminal and to the modem. Note the similarity between the Staco-to-modem and the Staco-to-terminal connections.

Example

If the terminal is in the SEND mode, the RTS line from terminal to Staco is active. When the Staco receives a poll, it activates the RTS line to the modem. After a predetermined delay, Staco receives a CTS from the modem. At this time, Staco activates the CTS to the terminal and the terminal starts transmitting information.

Asynchronous systems are used for small operations. Each terminal generally has only a keyboard, display and a printer. Terminals cannot communicate with each other directly, but the systems normally have a *store and forward* capability. A remote terminal, when polled, can send a message to the primary. As part of the message, the terminal specifies that the message is not intended for the primary but for some other terminal. The primary would store the message and, after selecting the designated terminal, would forward the message on to that terminal.

Data Link Control Characters

These control codes are entered from the keyboard during data preparation and by the LCU at the primary.

SOH (Start of Heading).   This code is optional. When used, it must precede the heading portion of the message. The heading may contain destination information, message numbers, date, time, priority, and security classification.

STX (Start of Text).   This code is entered immediately after the heading

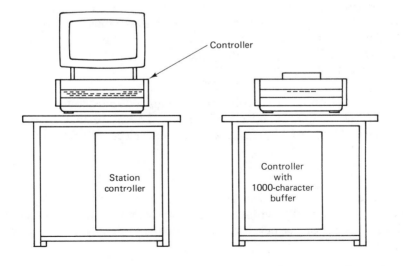

Figure 10–2 Block diagram of terminal equipment. (Courtesy of AT&T.)

Figure 10–3   Staco interface. (Courtesy of AT&T).

information and defines the end of the heading and the start of the text. In normal operation, the Stacos monitor the data link looking for their TSC or CDC. If text is being transmitted between the LCU and a previously polled/selected station, the text contents may inadvertently contain a bit sequence that represents the TSC or CDC of a third station. If this third station now started to transmit a message or an acknowledgment, two stations would be transmitting simultaneously and communications would be disrupted. There is a need for all Stacos not previously selected to ignore all text contents. When a Staco not previously selected or polled detects a STX or a SOH on the data link, it enters a "blind" mode—it ignores all transmitted characters until it detects an EOT. At this time, the Staco will "unblind" itself and again monitor the data link for its CDC or TSC.

EOT (End of Transmission).   This code must follow the last text character. Receipt of an EOT deselects all Stacos. To ensure that all remote stations are "unblinded," the LCU prefixes all polls and selections with EOT.

System Responses

Responses to polls and selections generally consist of two ASCII characters. The system used normally permits the user to program whatever characters he wishes for this function. For our purpose, we will define these characters as follows:

- Staco response to a poll:
  1. If the terminal is in the SEND mode, the message would be sent. Otherwise:

2. \ ACK   No Traffic, Ready to Receive.

3. \ \   No Traffic, Not Ready to Receive.

- Staco response to a selection:

1. \ ACK   Ready to Receive.

2. \ \   Not Ready to Receive. Terminal could be in LOCAL or out of service; the printer could be out of paper.

3. **   Not Ready to Receive, Traffic to Send.

- Optional Staco responses to a received message from the LCU (the station is capable of making these responses without being selected):

1. \ ACK   Entire message received with no parity error.

2. \ \   Entire message not received.

3. **   Station received the message with at least one parity error.

Dialogue

In addition to line protocol, means need to be established by which transmitted text can be positioned properly on the CRT. Table 10–1 lists two-character escape sequences that can be entered from the keyboard, transmitted as part of the text sequence, and are designed to cause specific actions at the terminal.

Example

Line	Col. 1
1	FORM 75B
6	NAME ************ AGE **
11	SEX * HT * FT ** IN
14	SHOE SIZE ***

This entire form is to be protected. All asterisks will be unprotected. They identify the locations where the operator may enter information. All unused parts of the form will be protected. Spaces shown on the form (i.e., after M of FORM) will also be protected.

```
E D                             LCU polls a station whose TSC is D.
O C D
T 3
```

```
S                     E         Station was in the SEND mode.
                 S
T F O R M    P  7 5 B O
X                     T          Requests Form 75B.
```

CPU would find this program among that station's applications program. After it was found:

```
E                               Station with the TSC of D has a CDC of BD.
O B D
T                                 That station is selected.
```

TABLE 10–1   ESCAPE SEQUENCE TERMINAL CONTROLS[a]

ESC sequence	Precedes transmission of	Reception of, unless elected not to
ESC 0	Each character where there is a tab setting	Sets tab stop at cursor location
ESC 1		Sets tab stop at cursor location, on that line and all lines below to end of memory
ESC 2		Clears all tab stops to right and below cursor location; stops to left and below remain, as do stops above
ESC 3	Each sequence of highlighted characters	Highlights characters that follow
ESC 4	Normal characters that follow highlighted characters	Does not highlight characters that follow
ESC W	Each sequence of protected characters	Protects characters that follow and puts display into protected data mode wherein receipt of on-line controls that effect unprotected data will also effect protected
ESC X	Normal characters that follow protected characters	Does not protect characters that follow and removes display from protected data mode wherein receipt of on-line controls that affect unprotected data will not affect protected data
ESC @		Moves cursor to first tab stop on right, or to start of next line if no tab stop on right, or to first unprotected space following protected data whether on that line or next
ESC 7		Moves cursor up one line
ESC B		Moves cursor down one line
ESC C		Moves cursor right one character
ESC G		Moves cursor to start of line
ESC H		Advances display to first segment and moves cursor to start of 1st line
ESC J		Clears data (unprotected only if not in protected data mode) to right of cursor, and all across all lines below cursor to end of memory
ESC L		Moves unprotected data down one line to create a line of spaces on the line the cursor is on and returns the cursor to the start of the line
ESC M		Clears unprotected data from the line the cursor is on, moves all unprotected data displayed below up one line, and returns the cursor to the start of the line
ESC ∧		Moves unprotected data one position to the right to create a character of space at the cursor position
ESC P		Clears unprotected character at the cursor position and moves all unprotected data

TABLE 10–1   ESCAPE SEQUENCE TERMINAL CONTROLS (*continued*)

ESC sequence	Precedes transmission of	Reception of, unless elected not to
		displayed on the right one position to the left
ESC R		Advances display to first segment, moves cursor to start of first line and clears all data from display memory, regardless of whether it is protected, unprotected, high-lighted, unhighlighted or tab stop
ESC S		Moves all displayed data up one line
ESC T		Moves all displayed data down one line
ESC U		Causes next whole segment to be displayed

[a](Courtesy of AT&T)

```
A                           Station is ready to receive.
\C
K
```

```
S E   D D E                               S
T S R E E S W F O R M S 7 5 B N N N N N N A M E S
X C   L L C           P       L L L L L           P
```

```
E                             E       E
S X * * * * * * * * * * * * * S W S A G E S S X * *
C                             C   P       P C
```

```
E                   E     E               E
S W N N N N N S E X S S X * S W S H T S S X *
C   L L L L L       P C   C   P     P C
```

```
E       S     E       E                   S       S
S W S F T S S X * * S W S I N N N N S H O E S I Z E S
C   P   P C       C     P   L L L       P           P
```

```
E           E                         E   E
S X * * * S W N N N N N N N N N N S @ O
C         C   L L L L L L L L L L C   T
```

Explanation of the transmitted form:

ESC R	Clears the screen and HOMEs the cursor.
Two DEL (or more)	Just a delay to allow the screen time to clear.
ESC W	The field that follows will be protected.

ESC X	The field that follows is unprotected. The operator can type information into these locations.
NL	New line. The last 10 NLs protect through line 24. For problem simplification, only one page of display memory is assumed.
ESC @	Places the cursor at the first unprotected space on the CRT. This is done as a courtesy to the operator.

QUESTIONS

1. An on-line terminal can be in one of three modes. What are the three modes?
2. When the operator at a remote site presses the SEND key, the message on the display is immediately transmitted. (T, F)
3. If the operator wants to type protected information on the display, which key must first be depressed?
4. What is meant by protected information?
5. If the remote terminal is solely printing out what is shown on the CRT, it is in the (*SEND, RECEIVE, LOCAL*) mode.
6. If a terminal is selected by the LCU, it is being asked to transmit if it has a message ready. (T, F)
7. If terminal is selected by the LCU while it is in the SEND mode, Staco (*sends "OK to transmit" to the primary, sends the terminal's message to the primary, sends "Not ready to receive, have a message to send" to the primary, does nothing to keep the primary in suspense*).
8. What is the normal line monitoring mode for the remote terminal?
9. Which two ASCII characters are capable of "blinding" all unselected stations?
10. Which ASCII character "unblinds" the station controllers?
11. A remote station can have more than one TSC. (T, F)
12. A remote station can have more than one CDC. (T, F)
13. Data link control characters may be entered directly from an asynchronous keyboard. (T, F)
14. Within a message, how is a field designated as protected?
15. If the remote terminal operator is composing a protected form, after the form is completed, how does the operator get out of the protected mode?

PROBLEMS

1. The operator at a remote station has requested Form 8561B from the primary. The format for this form is shown below. All characters included in the form are protected and all asterisks are unprotected. Write the character sequence that would display this form on the operator's CRT. Numbers under characters indicate the column the character is in.

```
LINE
  1        F O R M   8 5 6 1  B
           1       4  6        10

  2


  3        N A M E   * . . . *  S / S   * * * - * * - * * * *
           1       4  6      3032      36                    46
  4


  5        S T R E E T   * . . . *
           1            8   37
  6


  7        C I T Y   * . . . *   S T A T E   *   *
           1        6   30 32       36 38 39
  8


  9        Z I P   * * * * *
           1   3   5       9
```

11

BINARY SYNCHRONOUS COMMUNICATIONS

Binary Synchronous Communications (BSC or Bisync) is a synchronous, character-oriented protocol. ASCII and EBCDIC are generally used as the transmission codes with this protocol. ASCII will be used in all examples of transmission dialogue. Bisync is a *character-oriented* protocol because ASCII characters are used for data link control functions. This is in contrast with Synchronous Data Link Control (SDLC), a *bit-oriented* protocol, which uses bits or groups of bits to perform the data link control functions. Bisync is a *synchronous* protocol because SYN characters are sent with each transmission from which the receive DTE can achieve character synchronization.

A picture of a typical synchronous keyboard is shown in Figure 11–1. Note the absence of data link control character keys that are present on an asynchronous keyboard. This keyboard has two PA keys and 12 PF keys. The function of these keys will be explained as the need arises.

DATA LINK CONTROL CHARACTERS

(If a particular character is not an ASCII character, the ASCII character used will be shown in parentheses.)

SYN—Synchronous Idle. This character is transmitted to establish and maintain character synchronization. Unless operating in the transparent text mode (see later part of this chapter), this character is automatically stripped by the station controller. In a multipoint system, information is transmitted in blocks of 256 characters. In point-to-point systems, the length of one block can be considerably longer. If the length of a single transmission is 1 s or longer, SYN characters are inserted

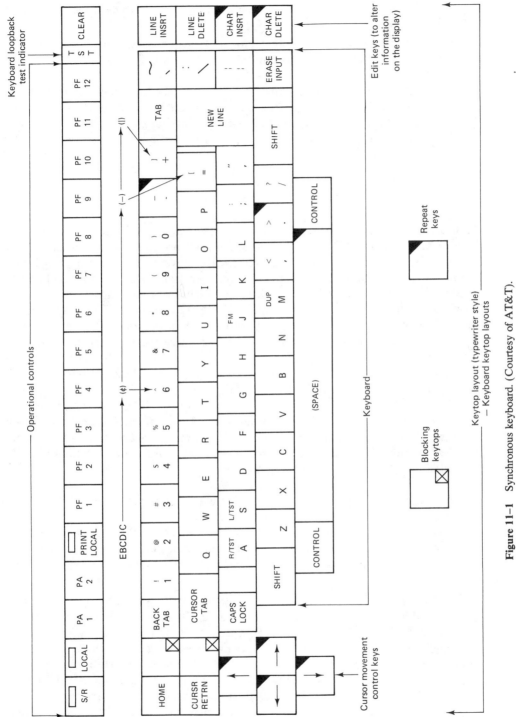

Figure 11-1 Synchronous keyboard. (Courtesy of AT&T).

171

directly into the message to maintain character synchronization. This would be necessary in some point-to-point transmissions.

EOT—End of Transmission.   This character indicates the end of the message and resets all stations on line to the control mode. In this mode, the secondaries are prepared to receive a poll or selection from the primary.

In a multipoint system, the primary starts all polls and selections with

```
P S S E P
A Y Y O A
D N N T D
```

The leading PAD is 55H or AAH (H = hexadecimal) and is used to establish bit synchronization. This PAD is not always used and will not be shown in any subsequent examples. Station controllers manufactured by IBM always supply this leading PAD. This is not true of station controllers built by other manufacturers. The SYN SYN (could be 2 or 5) pattern establishes character synchronization, and hence the name Bisync by which this protocol is often called. The EOT does not indicate end of transmission, but ensures that all secondaries are in the control mode. The trailing PAD (1 to 7) is FFH. It follows the last character of every transmission to allow that character to be demodulated by the receive modem and be sent to the DTE before the modem receiver turns off (see action of RLSD and RD of the RS 232C interface). All transmissions start with the SYN SYN pattern.

ETB—End of Transmission Block.   In a multipoint environment, a message may be more than one block long. ETB terminates the current block but infers that there are additional blocks to the message which will follow. The final block is terminated with ETX and not ETB or EOT.

ETX—End of Text.   This character is used to indicate the end of the current message.

ETB and ETX are always immediately followed by a BCC. A BCC must be *immediately acknowledged* by the receiver (except after ITB). Therefore, ETB BCC or ETX BCC always produce *line turnaround* (the station that was receiving will now transmit).

ITB—(US—Unit Separator)—End of Intermediate Block.   The function of this control character is similar to ETB. Although ITB must also be followed by a BCC, it differs from ETB in that its BCC will not be acknowledged immediately. The transmission of ITB will not produce a line turnaround. Eventually, a following block will be terminated by ETX. The receiving station must now acknowledge the BCCs. A positive acknowledgment indicates that the BCC following the ETB or ETX and all of the BCCs that followed previous ITBs are correct. If an error was detected in any of these BCCs, retransmission of all blocks will be required since the receiver has no way of informing the transmitter which specific BCCs were bad. The use of ITB speeds up message transmittal by eliminating line turnarounds required for acknowledgments. If, however, only one block is bad, all blocks must be retransmitted. ITB is used more in point-to-point systems where a single message may be quite lengthy. In a multipoint system, the length of a transmission is generally limited by the capacity of the terminal buffers.

ACK 0 (DLE 0), ACK 1 (DLE 1)—Positive Acknowledgments. These are positive acknowledgments of BCCs and are sent alternately. ACK 1 is always the positive acknowledgment to the first and odd blocks of text (all text blocks start with STX); ACK 0 is used to acknowledge even blocks of text and selections. A positive acknowledgment merely indicates no error in the BCC; it does not mean acceptance or compliance with the contents of the message.

WACK (DLE;)—Wait, Positive Acknowledgments. This character is used to provide positive acknowledgment to a BCC; it also requests the primary to delay further transmission to this station or device. A WACK is never sent by the primary to the secondary.

NAK—Negative Acknowledgment. This response indicates an error was detected in the BCC and that retransmission of that block (see ITB) is requested.

ENQ—Enquiry (format character). It is transmitted by the primary to indicate line turnaround. The secondary must transmit in response.

SOH—Start of Heading. SOH is generally followed by auxiliary information such as priority, time of message, and so on. SOH % R is used to begin a *Status and Sense message;* SOH % / is used for *Remote Test Request messages.*

STX—Start of Text. STX terminates a heading (SOH) if present and always precedes a block of text characters.

RVI—(DLE <)—Reverse Interrupt. This will be explained later.

MESSAGE FORMAT

The following information applies to multipoint, dedicated systems (Figure 11–2) which use Bisync protocol. Although there are systems that have as many as 50 devices controlled by a single station controller, the system under discussion will be considered to have a maximum of 32 devices that can be handled by a single station controller.

General Poll

Format

```
S S E P S S S    E P
Y Y O A Y Y P P " "  N A
N N T D N N A A    Q D
```

Courtesy of AT&T. SPA: Station Polling Address (Table 11–1).

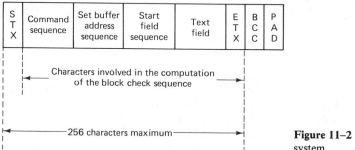

Figure 11–2 Multipoint dedicated system.

A general poll will always have two "characters following the SPA. A *general poll* is an invitation from the CPU to a particular station for any device at that station to send a message to the CPU. As the various devices at that station enter the SEND mode, the station controller keeps track of this sequence and will allow transmission in this order. In a general poll, the station controller determines which device will transmit if more than one are in the SEND mode.

Example

```
S S E P S S        E P
Y Y O A Y Y D D " "  N A
N N T D N N        Q D
```

General poll of station 4.

Specific Poll

Format

```
S S E P S S S S      E P
Y Y O A Y Y P P  D D  N A
                 A A
N N T D N N A A      Q D
```

Courtesy of AT&T. DA: Device Address (Table 11–1).

A *specific poll* is an invitation by the CPU to a particular device at a particular station to transmit. The CPU may send a specific poll in place of a general poll if there is only a small cluster of devices at that station. A specific poll of a device is required when the CPU wants a Status and Sense message from that device.

Example

```
S S E P S S        E P
Y Y O A Y Y D D H H  N A
N N T D N N        Q D
```

Specific poll of device 8 at station 4.

TABLE 11–1 STATION AND DEVICE ADDRESSES[a]

Station or device number	SPA	SSA	DA	Station or device number	SPA	SSA	DA
0	sp	-	sp	16	&	0	&
1	A	/	A	17	J	1	J
2	B	S	B	18	K	2	K
3	C	T	C	19	L	3	L
4	D	U	D	20	M	4	M
5	E	V	E	21	N	5	N
6	F	W	F	22	O	6	O
7	G	X	G	23	P	7	P
8	H	Y	H	24	Q	8	Q
9	I	Z	I	25	R	9	R
10	[	ǀ	[	26	]	:	]
11	.	,	.	27	$	#	$
12	<	%	<	28	*	@	*
13	(	—	(	29	)	'	)
14	+	>	+	30	;	=	;
15	!	?	!	31	∧	"	∧

[a](Courtesy of AT&T).

Selection

Format

```
S S E P S S S S     E P
Y Y O A Y Y S S D D N A
                A A
N N T D N N A A     Q D
```

Courtesy of AT&T. SSA: Station Selection Address (Table 11–1).

A *selection* means the CPU has a message for the device addressed and wants to determine if that device is in position to receive a message (i.e., not in local).

Example

```
S S E P S S         E P
Y Y O A Y Y T T F F N A
N N T D N N         Q D
```

Selection of device 6 at station 3.

No BCC is transmitted with a poll or selection. This necessitates the repetition of SPA, DA, and SSA. If an error occurred with the first SSA, it is unlikely that exactly the same error will occur with the second SSA. Unless both SSAs are identical, no station will respond. A retransmission of the poll or selection, hopefully, will not suffer the same problem.

If there is no response to a general poll (for any reason), the CPU waits 4 s and then retransmits the general poll. If no response, after another 4-s delay, a third general poll is sent. If there is still no response, the CPU will continue with its general poll of other stations. When it is time for the "problem" station to be polled, the CPU will again try three times at 4-s intervals. After the CPU has polled the station on 10 occasions, three times on each occasion at 4-s intervals, and it still has not raised a response, the CPU will drop that station from the poll sequence. Manual intervention will be required to replace that station on the poll sequence. The polling options are established at the CPU through software.

Block Check Characters (BCC).   The characters following the first SOH or STX in a block are involved in the computation of the BCC. The last character of a block involved in this computation is either ETB, ETX, or ITB.

Example

```
S  S  S        S            E  B  P
Y  Y  O        T            T  C  A
N  N  H |      X            X| C  D
        |_____|

S  S  S                     E  B  P
Y  Y  T                     T  C  A
N  N  X |                   B| C  D
        |_____|
```

Block characters involved in the computation of the BCC are shown grouped.

Message Format

1. *Commands*
   (a) Write:

```
E     W
S  1  C      (see Table 11–2 for WCC)
C     C
```

Will overwrite previous information in the data buffers with received information.

   (b) Erase-Write:

```
E     W
S  5  C
C     C
```

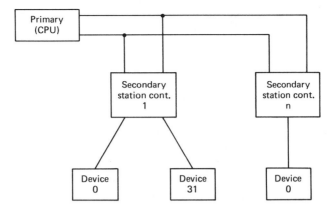

Figure 11–3 Multipoint dedicated system. (Courtesy of AT&T.)

TABLE 11–2 WRITE CONTROL CHARACTERS[a]

Start printer	Sound KD alarm	Put KD in LOCAL	Reset ACs to unmod	WCC			
				NL char/line	40 char/line	64 char/line	80 char/line
				sp	&	–	0
			×	A	J	/	1
		×		B	K	S	2
		×	×	C	L	T	3
	×			D	M	U	4
	×		×	E	N	V	5
	×	×		F	O	W	6
	×	×	×	G	P	X	7
×				H	Q	Y	8
×			×	I	R	Z	9
×		×		[¢	] !	¦	:
×		×	×	.	$	,	#
×	×			<	*	%	@
×	×		×	(	)	—	'
×	×	×		+	;	>	=
×	×	×	×	!	∧⌐	?	"

[a](Courtesy of AT&T).

Clears all data buffers before entering received data. Also homes the cursor.

(c) Erase All Unprotected:

```
E
S ?
C
```

Clears all unprotected fields in the buffer.

(d) Copy:

```
E   C D
S 7 C A    (see Table 11–7 for CCC)
C   C
```

Transfers specified contents of the buffer of a specified device into the buffer of the selected device on the same station cluster.

(e) Read Modified:

```
E
S 6
C
```

Transmits data fields flagged as modified from the buffer.

(f) Read All:

```
E
S 2
C
```

Transmits the contents of the buffer, including nulls.

WCC—Write Control Character.   A single control character chosen to supply the required conditions.

Example

WCC = 6

Conditions: Sounds KD alarm and puts KD in LOCAL and prints with 80 characters/line.

CCC—Copy Control Character (Table 11–7).   A single character identifying the type of information to be copied. The DA following this character identifies the device that will be copied from.

Example

CCC = 9

TABLE 11-3 ADDRESS MATRIX[a]

	B	B	C	C	A	A
	A	A			A	A
	1	2	1	2	1	2

Column

Row	1	2	3	4	5	6	7	8	9	10	11	12	13	14	15	16
1	S_P^S	S_P^A	S_P^B	S_P^C	S_P^D	S_P^E	S_P^F	S_P^G	S_P^H	S_P^I	$S_P^[$	$S_P^.$	$S_P^<$	$S_P^($	S_P^+	S, / P'
2	A&	AJ	AK	AL	AM	AN	AO	AP	AQ	AR	A]	A$	A*	A)	A;	A∧
3	B–	B/	BS	BT	BU	BV	BW	BX	BY	BZ	B;	B,	B%	B–	B>	B?
4	C0	C1	C2	C3	C4	C5	C6	C7	C8	C9	C:	C#	C@	C'	C=	C"
5	E_P^S	EA	EB	EC	ED	EE	EF	EG	EH	EI	E[	E.	E<	E(	E+	E!
6	F&	FJ	FK	FL	FM	FN	FO	FP	FQ	FR	F]	F$	F*	F)	F;	F∧
7	G–	G/	GS	GT	GU	GV	GW	GX	GY	GZ	G;	G,	G%	G–	G>	G?
8	H0	H1	H2	H3	H4	H5	H6	H7	H8	H9	H:	H#	H@	H'	H=	H"
9	$[_P^S$	[A	[B	[C	[D	[E	[F	[G	[H	[I	[[	[.	[<	[(	[+	[!
10	.&	.J	.K	.L	.M	.N	.O	.P	.Q	.R	.]	.$	.*	.)	.;	.∧
11	<–	</	<S	<T	<U	<V	<W	<X	<Y	<Z	<;	<,	<%	<–	<>	<?
12	(0	(1	(2	(3	(4	(5	(6	(7	(8	(9	(:	(#	(@	('	(=	("
13	$!_P^S$	!A	!B	!C	!D	!E	!F	!G	!H	!I	![	!.	!<	!(	!+	!!
14	&&	&J	&K	&L	&M	&N	&O	&P	&Q	&R	&]	&$	&*	&)	&;	&∧
15	J–	J/	JS	JT	JU	JV	JW	JX	JY	JZ	J;	J,	J%	J–	J>	J?
16	K0	K1	K2	K3	K4	K5	K6	K7	K8	K9	K:	K#	K@	K'	K=	K"
17	M_P^S	MA	MB	MC	MD	ME	MF	MG	MH	MI	M[	M.	M<	M(	M+	M!
18	N&	NJ	NK	NL	NM	NN	NO	NP	NQ	NR	N]	N$	N*	N)	N;	N∧
19	O–	O/	OS	OT	OU	OV	OW	OX	OY	OZ	O;	O,	O%	O–	O>	O?
20	P0	P1	P2	P3	P4	P5	P6	P7	P8	P9	P:	P#	P@	P'	P=	P"
21	R_P^S	RA	RB	RC	RD	RE	RF	RG	RH	RI	R[	R.	R<	R(	R+	R!
22	]&	]J	]K	]L	]M	]N	]O	]P	]Q	]R	]]	]$	]*	])	];	]∧
23	$–	$/	$S	$T	$U	$V	$W	$X	$Y	$Z	$;	$,	$%	$–	$>	$?
24	*0	*1	*2	*3	*4	*5	*6	*7	*8	*9	*:	*#	*@	*'	*=	*"

[a](Courtesy of AT&T).

TABLE 11-3 (*continued*)

B B C C A A
A A A A A A
1 2 1 2 1 2

Column

Row	17	18	19	20	21	22	23	24	25	26	27	28	29	3C	31	32
1	S&/P	SJ/P	SK/P	SL/P	SM/P	SN/P	SO/P	SP/P	SQ/P	SR/P	S]/P	S$/P	S*/P	S)/P	S;/P	S∧/P
2	A−	A/	AS	AT	AU	AV	AW	AX	AY	AZ	A;	A,	A%	A−	A>	A?
3	B0	B1	B2	B3	B4	B5	B6	B7	B8	B9	B:	B#	B@	B'	B=	B"
4	DS/DP	DA	DB	DC	DD	DE	DF	DG	DH	DI	D[	D.	D<	D(	D+	D!
5	E&	EJ	EK	EL	EM	EN	EO	EP	EQ	ER	E]	E$	E*	E)	E;	E∧
6	F−	F/	FS	FT	FU	FV	FW	FX	FY	FZ	F;	F,	F%	F_	F>	F?
7	G0	G1	G2	G3	G4	G5	G6	G7	G8	G9	G:	G#	G@	G'	G=	G"
8	IS/1P	IA	IB	IC	ID	IE	IF	IG	IH	II	I[	I.	I<	I(	I+	I!
9	[&	[J	[K	[L	[M	[N	[O	[P	[Q	[R	[]	[$	[*	[)	[:	[<
10	.−	./	.S	.T	.U	.V	.W	.X	.Y	.Z	.:	.:	.%	.!	.>	.?
11	<0	<1	<2	<3	<4	<5	<6	<7	<8	<9	<:	<#	<@	<'	<=	<"
12	+S/+P	+A	+B	+C	+D	+E	+F	+G	+H	+I	+[	+.	+<	+(	++	+!
13	!&	!J	!K	!L	!M	!N	!O	!P	!Q	!R	!]	!$	!*	!)	!;	!∧
14	&−	&/	&S	&T	&U	&V	&W	&X	&Y	&Z	&;	&,	&%	&−	&>	&?
15	J0	J1	J2	J3	J4	J5	J6	J7	J8	J9	J:	J#	J@	J'	J=	J"
16	LS/LP	LA	LB	LC	LD	LE	LF	LG	LH	LI	L[	L.	L<	L(	L+	L!
17	M&	MJ	MK	ML	MM	MN	MO	MP	MQ	MR	M]	M$	M*	M)	M;	M∧
18	N−	N/	NS	NT	NU	NV	NW	NX	NY	NZ	N;	N,	N%	N_	N>	N?
19	O0	O1	O2	O3	O4	O5	O6	O7	O8	O9	O:	O#	O@	O'	O=	O"
20	QS/QP	QA	QB	QC	QD	QE	QF	QG	QH	QI	Q[	Q.	Q<	Q(	Q+	Q!
21	R&	RJ	RK	RL	RM	RN	RO	RP	RQ	RR	R]	R$	R*	R)	R;	R∧
22	]−	]/	]S	]T	]U	]V	]W	]X	]Y	]Z	];	],	]%	]_	]>	]?
23	$0	$1	$2	$3	$4	$5	$6	$7	$8	$9	$:	$#	$@	$'	$=	$"
24	)S/)P	)A	)B	)C	)D	)E	)F	)G	)H	)I	)[	).	)<	)(	)+	)!

TABLE 11-3 (continued)

B	B	C	C	A	A
A	A	A	A	A	A
1	2	1	2	1	2

Column

Row	33	34	35	36	37	38	39	40	41	42	43	44	45	46	47	48
1	S_- P_-	$S_/$ P'	S_S P'	S_T P'	S_U P'	S_V P'	S_W P'	S_X P'	S_Y P'	S_Z P'	$S_;$ P'	$S_,$ P'	$S_\%$ P'	S P_-	$S_>$ P'	$S_?$ P'
2	A0	A1	A2	A3	A4	A5	A6	A7	A8	A9	A:	A#	A@	A'	A=	A"
3	C_P^S	CA	CB	CC	CD	CE	CF	CG	CH	CI	C[	C.	C<	C(	C+	C!
4	D&	DJ	DK	DL	DM	DN	DO	DP	DQ	DR	D]	D$	D*	D)	D;	D\
5	E−	E/	ES	ET	EU	EV	EW	EX	EY	EZ	E!	E,	E%	E−	E>	E?
6	F0	F1	F2	F3	F4	F5	F6	F7	F8	F9	F:	F#	F@	F'	F=	F"
7	H_P^S	HA	HB	HC	HD	HE	HF	HG	HH	HI	H[	H.	H<	H(	H+	H!
8	I&	IJ	IK	IL	IM	IN	IO	IP	IQ	IR	I]	I$	I*	I)	I;	I\
9	[−	[/	[S	[T	[U	[V	[W	[X	[Y	[Z	[!	[,	[%	[)	[>	[?
10	.0	.1	.2	.3	.4	.5	.6	.7	.8	.9	.:	.#	.@	.,	.=	.'
11	$(_P^S$	(A	(B	(C	(D	(E	(F	(G	(H	(I	([	(.	(<	((	(+	(!
12	+&	+J	+K	+L	+M	+N	+O	+P	+Q	+R	+]	+$	+*	+)	+;	+\
13	!−	!/	!S	!T	!U	!V	!W	!X	!Y	!Z	!:	!,	!%	!'	!>	!?
14	&0	&1	&2	&3	&4	&5	&6	&7	&8	&9	&:	&#	&@	&'	&=	&"
15	K_P^S	KA	KB	KC	KD	KE	KF	KG	KH	KI	K[	K.	K<	K(	K+	K!
16	L&	LJ	LK	LL	LM	LN	LO	LP	LQ	LR	L]	L$	L*	L)	L;	L\
17	M−	M/	MS	MT	MU	MV	MW	MX	MY	MZ	M;	M,	M%	M_	M>	M?
18	N0	N1	N2	N3	N4	N5	N6	N7	N8	N9	N:	N#	N@	N'	N=	N"
19	P_P^S	PA	PB	PC	PD	PE	PF	PG	PH	PI	P[	P.	P<	P(	P+	P!
20	Q&	QJ	QK	QL	QM	QN	QO	QP	QQ	QR	Q]	Q$	Q*	Q)	Q;	Q\
21	R−	R/	RS	RT	RU	RV	RW	RX	RY	RZ	R!	R,	R%	R_	R>	R?
22	]0	]1	]2	]3	]4	]5	]6	]7	]8	]9	]:	]#	]@	]'	]=	]"
23	$*_P^S$	*A	*B	*C	*D	*E	*F	*G	*H	*I	*[	*.	*<	*(	*+	*!
24	)&	)J	)K	)L	)M	)N	)O	)P	)Q	)R	)]	)$	)*	))	);	)\

181

TABLE 11–3 (continued)

B	B	C	C	A	A
A	A	A	A	A	A
1	2	1	2	1	2

Column

Row	49	50	51	52	53	54	55	56	57	58	59	60	61	62	63	64
1	$\frac{S}{P^0}$	$\frac{S_1}{P^1}$	$\frac{S_2}{P^2}$	$\frac{S_3}{P^3}$	$\frac{S_4}{P^4}$	$\frac{S_5}{P^5}$	$\frac{S_6}{P^6}$	$\frac{S_7}{P^7}$	$\frac{S_8}{P^8}$	$\frac{S_9}{P^9}$	$\frac{S:}{P:}$	$\frac{S\#}{P\#}$	$\frac{S@}{P@}$	$\frac{S,}{P,}$	$\frac{S=}{P=}$	$\frac{S"}{P"}$
2	$\frac{BS}{P}$	BA	BB	BC	BD	BE	BF	BG	BH	BI	B|	B.	B<	B(	B+	B!
3	C&	CJ	CK	CL	CM	CN	CO	CP	CQ	CR	C|	C$	C*	C)	C;	C∧
4	D−	D/	DS	DT	DU	DV	DW	DX	DY	DZ	D|	D,	D%	D_	D>	D?
5	E0	E1	E2	E3	E4	E5	E6	E7	E8	E9	E:	E#	E@	E'	E=	E"
6	$\frac{GS}{P}$	GA	GB	GC	GD	GE	GF	GG	GH	GI	G|	G.	G<	G(	G+	G!
7	H&	HJ	HK	HL	HM	HN	HO	HP	HQ	HR	H|	H$	H*	H)	H;	H∧
8	I−	I/	IS	IT	IU	IV	IW	IX	IY	IZ	I|	I,	I%	I_	I>	I?
9	|0	|1	|2	|3	|4	|5	|6	|7	|8	|9	|:	|#	|@	|'	|=	|"
10	$\frac{<S}{P}$	<A	<B	<C	<D	<E	<F	<G	<H	<I	<|	<.	<<	<(	<+	<!
11	(&	(J	(K	(L	(M	(N	(O	(P	(Q	(R	(|	($	(*	()	(;	(∧
12	+−	+/	+S	+T	+U	+V	+W	+X	+Y	+Z	+|	+,	+%	+_	+>	+?
13	!0	!1	!2	!3 \	!4	!5	!6	!7	!8	!9	!:	!#	!@	!'	!=	!"
14	$\frac{JS}{P}$	JA	JB	JC	JD	JE	JF	JG	JH	JI	J|	J.	J<	J(	J+	J!
15	K&	KJ	KK	KL	KM	KN	KO	KP	KQ	KR	K|	K$	K*	K)	K;	K∧
16	L−	L/	LS	LT	LU	LV	LW	LX	LY	LZ	L|	L,	L%	L_	L>	L?
17	M0	M1	M2	M3	M4	M5	M6	M7	M8	M9	M:	M#	M@	M'	M=	M"
18	$\frac{OS}{P}$	OA	OB	OC	OD	OE	OF	OG	OH	OI	O|	O.	O<	O(	O+	O!
19	P&	PJ	PK	PL	PM	PN	PO	PP	PQ	PR	P|	P$	P*	P)	P;	P∧
20	Q−	Q/	QS	QT	QU	QV	QW	QX	QY	QZ	Q|	Q.	Q%	Q_	Q>	Q?
21	R0	R1	R2	R3	R4	R5	R6	R7	R8	R9	R:	R#	R@	R'	R=	R"
22	$\frac{\$S}{P}$	$A	$B	$C	$D	$E	$F	$G	$H	$I	$|	$.	$<	$(	$+	$!
23	*&	*J	*K	*L	*M	*N	*O	*P	*Q	*R	*|	*$	**	*)	*;	*∧
24	)−	)/	)S	)T	)U	)V	)W	)X	)Y	)Z	):	),	)%	)_	)>	)?

TABLE 11-3 (continued)

```
B B  C C  A A
A A  A A  A A
1 2  1 2  1 2
```

Column

Row	65	66	67	68	69	70	71	72	73	74	75	76	77	78	79	80
1	A S P	AA	AB	AC	AD	AE	AF	AG	AH	AI	A[	A.	A<	A(	A+	A!
2	B&	BJ	BK	BL	BM	BN	BO	BP	BQ	BR	B]	B$	B*	B)	B;	B\
3	C-	C/	CS	CT	CU	CV	CW	CX	CY	CZ	C¦	C,	C%	C_	C>	C?
4	D0	D1	D2	D3	D4	D5	D6	D7	D8	D9	D:	D#	D@	D'	D=	D"
5	F S P	FA	FB	FC	FD	FE	FF	FG	FH	FI	F[	F.	F<	F(	F+	F!
6	G&	GJ	GK	GL	GM	GN	GO	GP	GQ	GR	G]	G$	G*	G)	G;	G\
7	H-	H/	HS	HT	HU	HV	HW	HX	HY	HZ	H¦	H,	H%	H_	H>	H?
8	I0	I1	I2	I3	I4	I5	I6	I7	I8	I9	I:	I#	I@	I'	I=	I"
9	. S P	.A	.B	.C	.D	.E	.F	.G	.H	.I	.[	..	.<	.(	.+	.!
10	<&	<J	<K	<L	<M	<N	<O	<P	<Q	<R	<]	<$	<*	<)	<;	<\
11	(-	(/	(S	(T	(U	(V	(W	(X	(Y	(Z	(¦	(,	(%	(_	(>	(?
12	+0	+1	+2	+3	+4	+5	+6	+7	+8	+9	+:	+#	+@	+'	+=	+"
13	& S P	&A	&B	&C	&D	&E	&F	&G	&H	&I	&[	&.	&<	&(	&+	&!
14	J&	JJ	JK	JL	JM	JN	JO	JP	JQ	JR	J]	J$	J*	J)	J;	J\
15	K-	K/	KS	KT	KU	KV	KW	KX	KY	KZ	K¦	K,	K%	K_	K>	K?
16	L0	L1	L2	L3	L4	L5	L6	L7	L8	L9	L:	L#	L@	L'	L=	L"
17	N S P	NA	NB	NC	ND	NE	NF	NG	NH	NI	N[	N.	N<	N(	N+	N!
18	O&	OJ	OK	OL	OM	ON	OO	OP	OQ	OR	O]	O$	O*	O)	O;	O\
19	P-	P/	PS	PT	PU	PV	PW	PX	PY	PZ	P¦	P,	P%	P_	P>	P?
20	Q0	Q1	Q2	Q3	Q4	Q5	Q6	Q7	Q8	Q9	Q:	Q#	Q@	Q'	Q=	Q"
21	| S P	|A	|B	|C	|D	|E	|F	|G	|H	|I	|[	|.	|<	|(	|+	|!
22	S&	SJ	SK	SL	SM	SN	SO	SP	SQ	SR	S]	S$	S*	S)	S;	S\
23	*-	*/	*S	*T	*U	*V	*W	*X	*Y	*Z	*¦	*,	*%	*_	*>	*?
24	)0	)1	)2	)3	)4	)5	)6	)7	)8	)9	):	)#	)@	)'	)=	)"

Want to start printer and copy only the unprotected data and ACs. Print at 80-character/line format.

2. *Device order:* Used by the line control unit (LCU) to format the message.
   (a) Set Buffer Address:

```
D B B
C A A      (see table 11–3 for BA1 and BA2)
1 1 2
```

Identifies the buffer address at which the operation will begin or continue. The buffer address is identified by two characters: BA1, BA2 (buffer addresses 1 and 2). These addresses identify the starting row and column.

Example

```
D          BA1 = F   BA2 = K
C F K
1          Identifies row 6, column 3
```

(b) Start Field:

```
G A
S C
```

Indicates the start of a new field. The AC (Attribute Character; see Table 11–4) immediately following the GS (ASCII Group Separator) identifies the characteristics of this field. The field will extend up to the location where the next AC is stored. On the screen, the field will wrap-around if an AC is stored in row 23, column 7 and no other AC exists below it (to and including row 24, column 80), the field continues to HOME and the top of the screen until the next AC is found.

Example

```
G Y        The following field is protected and intensified.
S
```

The GS simply indicates that the next received character is an AC. When received, the GS is discarded. The AC is stored in a *protected* character position in the buffer but is displayed on the screen as a *space*.

AC: The contents of the device buffers are broken up into *fields.* Each field starts with a field AC which denotes the properties of that field.

The types of fields described by attribute characters are as follows:

(1) *Protected:* accessible only by the remote LCU. The terminal operator cannot enter or alter data in this field.

(2) *Unprotected:* accessible to the operator and LCU.

TABLE 11–4 ATTRIBUTE CHARACTERS[a]

AC	Protected	Numeric	Hidden	Blinked	Intensified	Modified
sp					X	X
A					X	X
H						X
I						X
J			X			
.				X		
∨				X		
(			X			
&		X				X
J		X				X
Q		X			X	
R		X			X	
]		X		X		
S		X		X		X
*		X	X			
)		X	X			X

AC	Protected	Numeric	Hidden	Blinked	Intensified	Modified
-	X				X	X
/	X				X	X
Y	X			X		X
Z	X			X		X
--	X		X			X
.	X		X			X
%	X					X
|	X					X
0	X	X				X
1	X	X			X	X
8	X	X		X	X	X
9	X	X		X		X
:	X	X				X
#	X	X				X
@	X	X	X			X
,	X	X	X			X

[a](Courtesy of AT&T).

(3) *Numeric:*   allows only 0–9, DUP, minus, and period to be entered in this field by the operator.

(4) *Hidden:*   data in this field are stored in the display buffers but are not visible on the screen or printed on the printer.

(5) *Modified:*   fields in which data have been altered by the operator or which has been declared as such by the AC.

Note: The write or erase-write command is followed by a WCC. A WCC may declare all ACs as unmodified. Normally, when data are transmitted, only modified data are sent (data which the operator has entered). This eliminates unnecessary transmission of a form and simply sends out the information that was entered in completing the form. When required, however, the data field of the form itself may be declared as modified. Then, when transmission takes place, both the form and the data required to complete the form are transmitted as modified data.

(6) *Intensified (highlighted):* characters in this field are displayed at a higher than normal intensity on the display (X 1.5).

(7) *Blinked:* characters in this field are displayed alternately between full and half intensity.

(8) *Protected and Numeric:* these two attributes combine to provide *AUTOSKIP.* Any time the cursor falls on an autoskip AC, the cursor automatically skips to the first location of the next unprotected field. Where conditions exist, it will wrap-around.

(c) Insert Cursor:

```
D     Places cursor at the current buffer address.
C
3
```

Example

```
D    D     Places cursor at row 6, column 3.
C F K C
1     3
```

(d) Program Tab:

```
            H
            T
```

Advances the buffer pointer to the first location of the next unprotected field. If currently on an attribute of an unprotected field, the cursor advances one position. If the pointer is in an unprotected field and follows either text or another HT, the rest of the field is replaced by nulls.

(e) Repeat to Address:

```
D B B
C A A C
4 1 2
```

Stores a specified character into every buffer position from the present position up to but excluding the specified buffer position. Will wrap-around. If the buffer address is the same as the current address, the character will be placed in all locations.

Example

```
D
C S N T
  P
4
```

Assuming that the current address was HOME, T would be placed in every location up to but not including row 1, column 22.

(f) Erase Unprotected to Address:

```
D B B     Self-explanatory.
C A A
2 1 2     Will wrap-around.
```

If the buffer address equals the current address, nulls are set in all unprotected locations.

AIDS—Attention Identifiers

Refer to the layout of the synchronous keyboard in Figure 11–1. When a PA or a PF key is depressed, the ASCII code corresponding to the AID (Table 11–5) of the depressed key is transmitted (when polled).

The PA (*Program Access*) keys request a preformatted form from the CPU. No text can be transmitted by the secondary when these keys are depressed.

The PF (*Program Form*) keys can cause a variety of options to occur: the CPU can transmit a preformatted form; the CPU can initiate a copy command; anything. Text can be transmitted with these AIDs.

Each AID is just a single character. When the CPU receives that character, it goes to the application programs for that station to determine the action necessitated by that character. Therefore, the function of each PF key depends on what has been programmed *for that station* and *for that key*.

Format for Secondary Messages

1. Read Modified Message (in response to poll or read modified command if PA or Clear key is depressed) (Courtesy of AT&T):

```
┌─────────────┐
│ S S    A E B │
│ T P D  I T C │
│    A         │
│ X A    D X C │
└─────────────┘
```

Example

Operator of device 1 at station 2 depressed the PA1 key. After being polled, the station would transmit:

```
S S S       E B P
Y Y T B A % T C A
N N X       X C D
```

2. Read All Message (in reply to a Read All command) (Courtesy of AT&T):

S S	A C C	G A	data in	E B
T P D	I A A	S C	fields +	T C
A			nulls	
X A	D 1 2			X C
			for each field	

Example

```
S S E P S S        E P
Y Y O A Y Y / / C C N A ─────────────→
N N T D N N         Q D
```

Selection of station 1, device 3: a KD.

```
         S S D   P
────────← Y Y L 0 A ─────────
         N N E   D
```

Positive acknowledgment; device 3 ready to receive a message.

```
         S S S E   E B P
──────→  Y Y T S 2 T C A ─────────→
         N N X C   X C D
```

Primary sends· a Read All command.

```
    S S S                              E B P
──← Y Y T / C ·  S S G A data, including nulls  T C A ─←
    N N X         P P S C   for each field       X C D
```

Secondary sends everything in its buffers.

Note that in Table 11–5, no key is depressed by the operator in response to a Read All command. This means that the primary could read the present contents of the secondary device's buffers without the operator at that station knowing about it. The AID character is inserted by the station controller. This command/response is useful in some troubleshooting situations.

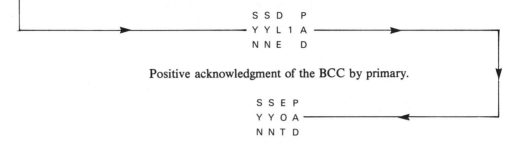

```
    S S D   P
    Y Y L 1 A
    N N E   D
```

Positive acknowledgment of the BCC by primary.

```
    S S E P
    Y Y O A
    N N T D
```

Handshake by the secondary.

TABLE 11–5 ATTENTION IDENTIFIERS (AIDs)[a]

Key depressed	AID	Message response to poll
S/R	'	Read Modified
PA1	%	Short Read
PA2	<	Short Read
PA3	,	Short Read
R/TST	0	Read Modified
PF1	1	Read Modified
PF2	2	Read Modified
PF3	3	Read Modified
PF4	4	Read Modified
PF5	5	Read Modified
PF6	6	Read Modified
PF7	7	Read Modified
PF8	8	Read Modified
PF9	9	Read Modified
PF10	.	Read Modified
PF11	#	Read Modified
PF12	@	Read Modified
CLEAR	___	Short Read
None, KD	-	Response to Read
None, P	Y }	All Command Only
Magnetic Stripe Reader	W	Read Modified

[a](Courtesy of AT&T).

3. Read Modified Message (in reply to a poll or to a Read Modified command if either the S/R or PF key is depressed) (Courtesy of AT&T).

S S	A C C	D A A data in	E B
T P ^D_A I A A	C A A fields	T C
X A D 1 2	1 1 2 less nulls	X C
	for each modified field	

Example

```
S  S E P  S S           E P
Y  Y O A Y  Y B B D D N A ──────────────────────►
N  N T D N  N          Q D
```

Specific poll of device 4 at station 2.

If the S/R key had been depressed by the operator:

```
       S  S S                D    data in field less   E B P
◄───── Y  Y T B D ' B T      C A J nulls for each      T C A ◄───
       N  N X                1    modified field       X C D
```

Cursor was left at row 3, column 4; first modified character is in row 2, column 2.

```
        S  S D  P
───────► Y  Y L 1 A ──────────────────────►
        N  N E  D
```

Primary sends positive acknowledgment of BCC.

```
        S  S E P
        Y  Y O A ──────
        N  N T D
```

Handshake by the secondary.

Points to keep in mind.

(a) All transmissions containing a BCC must be acknowledged. BCCs after an ETB or ETX must be acknowledged immediately.

(b) Negative acknowledgments require retransmission of the last block received.

(c) If the secondary is sending the message, the secondary will handshake the primary with

```
        S  S E P
        Y  Y O A
        N  N T D
```

after it receives a final positive acknowledgment of its BCC (after ETX). If the primary is sending the message, it does not handshake the secondary after it receives a final positive acknowledgment.

(d) The secondary cannot transmit or receive unless it is first polled or selected, respectively, by the primary.

(e) Although there is an ASCII code for ACK, DLE is used as the transmission code for this purpose.

(f) In the transmission of certain forms, the buffer address (BA1 and BA2), the cursor address (CA1 and CA2), or the attribute address (AA1 and AA2) need to be specified. The same coding is used for all of these addresses. The buffer address is the specified location in memory; the cursor address is the current address of the cursor. However, where an attribute address is specified, it does not mean the address where the attribute character is stored but the *first location following this address.*

Example

Given:

Assume that the attribute Y is stored in row 3, column 5. If the attribute address was called for by any format, the address for row 3, column 6 should be given.

(g) *Difference between NULL and SPACE:* If a device's memory is erased, NULLS (ASCII = 00H) are stored in all locations and displayed on the screen as blank spaces. A SPACE (ASCII = 20H) is considered a separation character. It would also be displayed on the screen as a blank space. During the transmission of modified data, if a SPACE was entered in an unprotected field (operator hit the space bar), it would be transmitted. A NULL in the same location would not be transmitted.

(h) A field is declared unprotected and unmodified. If it is unprotected, the operator can write into that field. Once the operator enters characters into that field, since he or she changed the characters from what they were, those characters are considered modified data.

Example

A form is already prepared and stored in computer memory at the primary. The secondary can obtain this form by depressing the PF6 key (this key was programmed for this function). Device 6 at station 2 is the secondary in question. The form is as follows:

C L A S S ∗(5 unprotected spaces) C O U R S E ∗(10 unprotected spaces).
(∗ denotes a protected space.)
The C of CLASS will be on row 5, column 20.
The C of COURSE will be on row 5, column 40.

When the form is sent:

Command: sound tone, reset attributes, 64 characters/line
Form: protected and intensified
Operator-entered information: blinked

Dialogue. PF6 key is depressed by the operator.

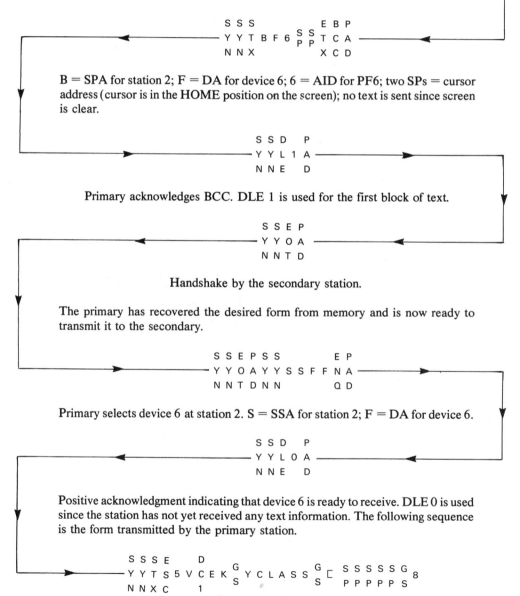

```
S S E P S S        E P
Y Y O A Y Y B B " "  N A  ─────────►
N N T D N N        Q D
```

General poll of station 2. Station controller detects that device 6 has its PF6 key depressed.

```
       S S S           E B P
  ◄─── Y Y T B F 6 S S T C A ───
       N N X     P P   X C D
```

B = SPA for station 2; F = DA for device 6; 6 = AID for PF6; two SPs = cursor address (cursor is in the HOME position on the screen); no text is sent since screen is clear.

```
     S S D  P
 ──► Y Y L 1 A ─────►
     N N E  D
```

Primary acknowledges BCC. DLE 1 is used for the first block of text.

```
     S S E P
 ◄── Y Y O A ───
     N N T D
```

Handshake by the secondary station.

The primary has recovered the desired form from memory and is now ready to transmit it to the secondary.

```
     S S E P S S          E P
 ──► Y Y O A Y Y S S F F N A ─────►
     N N T D N N          Q D
```

Primary selects device 6 at station 2. S = SSA for station 2; F = DA for device 6.

```
     S S D  P
 ◄── Y Y L 0 A ───
     N N E  D
```

Positive acknowledgment indicating that device 6 is ready to receive. DLE 0 is used since the station has not yet received any text information. The following sequence is the form transmitted by the primary station.

```
     S S S E    D             G          G    S S S S S G
 ──► Y Y T S 5 V C E K S Y C L A S S S [ S   P P P P P S 8
     N N X C    1
```

```
          D                    D        D      D E B P
C E X C O U R S E  G   ⌐  C E 8  G   8 C E R C T C A ——————————————→
          1        S      1      S      1      3 X C D
```

Explanation of message:

ESC 5	Erase-write command. Clears screen, homes cursor.
V	WCC. Sounds KD alarm, resets ACs to unmodified, 64 characters/line.
DC1	Device order. The next two characters will represent a buffer address.
EK	Row 5, column 19.
GS	Indicated field will start at row 5, column 19. The type of field will be specified by the attribute character that follows.
Y	Attribute character. Field is protected and intensified. Remember, the attribute character is displayed as a protected space. Row 5, column 19 is now a protected space.
CLASS	This is the text. C is stored in row 5, column 20 and the last S will be in row 5, column 24.
GS	Indicated field will start at the next location—row 5, column 25.
⌐	Attribute character. Characters entered in this field will be blinked. The attribute character will also provide the protected space after CLASS.
5 SPs	Spaces allowed for the operator to enter his information.
GS	Indicated field will start at row 5, column 31.
8	Attribute character. Field is protected, numeric, and intensified. Protected and numeric = autoskip. When the operator enters the fifth character after CLASS, the cursor will move to the next location, which is an autoskip field. Once the cursor finds itself in this field, it moves on to the first unprotected location of the next unprotected field. The operator would then be automatically ready to enter the course.
DC1	Set buffer address to ...
EX	Row 5, column 40.
COURSE	Text that would start at row 5, column 40. The E ends up on row 5, column 45. COURSE is being written in a field identified as protected and numeric. The field is protected only as far as the device operator is concerned. The operator cannot overwrite the word COURSE. In fact, the operator cannot enter anything in this field. Similarly, numeric applies only to what the operator may enter. COURSE is not numeric, but the operator does not enter COURSE; it is part of the form. If the field is protected, it is entirely incidental whether the field is numeric or not as far as the operator is concerned; the operator cannot enter anything into this field anyway.
GS	Indicated field will start at the next location: row 5, column 46.
⌐	Attribute character. Field is blinked. Row 5, column 46 is a protected space for the attribute character. This also provides the required protected space after COURSE.
DC1	Set buffer address to ...
E8	Row 5, column 57.
GS	New field starts at row 5, column 57. This means that the operator can

	enter information (which will be blinked) from row 5, column 47 to row 5, column 56, which are the 10 required spaces. This is shown as an alternative method for providing these spaces in contrast with the way this was achieved after CLASS.
8	Attribute character. Protected, numeric, and intensified. Since there is no further form information, this protects the rest of the form. Remember, protected will wrap-around, so this AC protects the form from row 5, column 57 to row 24, column 80 and then back around from row 1, column 1 to row 5, column 18. Also, since this is an autoskip field, if the operator entered the tenth character in the preceding blinked field, the cursor would move into this autoskip field and end up at row 5, column 26. If programmed correctly, often the CPU can determine which form was completed simply from the final position of the cursor.
DC1	Set buffer address to . . .
ER	Row 5, column 26.
DC3	Insert cursor. This is done as a courtesy to the operator so that he or she may start completing the form without any unnecessary waste of time.

Station acknowledges BCC. Primary goes on to poll station 3.

(i) *Difference between unprotected and modified:* An unprotected field is a field in which the operator can enter data. In the previous message form, 10 spaces were left unprotected after COURSE for operator data. If the operator entered MATH, these four spaces immediately become modified: the contents of those locations were changed by the operator.

After entering MATH, if the operator hit the TAB key, the cursor would automatically move to the first unprotected space of the next field. The remaining spaces in the previous field would contain nulls which were not modified. If only modified data were transmitted, only MATH in the field after COURSE would be transmitted.

After entering MATH, if the operator repeatedly hit the space bar to get to the first unprotected space of the next field, the character for SPACE would have been entered in the six locations following MATH and the contents of these six locations would also have become modified data. If only modified data were transmitted, MATH and six 20H characters (20H = ASCII space) would be transmitted.

What if the attribute character after COURSE defined the next field to be modified and unprotected? If only MATH was entered by the operator and then he hit the TAB key, MATH, and six NULLS would be transmitted as modified data. Although the operator changed only four characters in this field, the entire field was declared as modified by the attribute character and the entire field would be transmitted whether the operator entered anything into it or not.

Similarly, if the COURSE field was declared protected and modified by the attribute character, the operator could not overwrite COURSE. However, COURSE would be transmitted whenever modified data were transmitted.

4. Status and Sense Message (Courtesy of AT&T):

```
S       S S     S S E B P
O % R T P D S S T C A
        X A  A  1 2 X C D
H
```

Transmission of RVI—Reverse Interrupt—by the receiving station constitutes a positive acknowledgment of a received BCC. In addition, the RVI requests the sending station to halt transmission at its earliest convenience because the receiving station has something to send of higher priority. This application of an RVI is used mainly in a two-point system where a block of data may be much larger than 256 characters. In a multipoint data link system, the length of a message from the primary is limited by the capacity of the terminal buffers at the secondary station. In a multipoint system, the RVI still serves as a positive acknowledgment, but it also indicates to the primary that the secondary is experiencing some form of difficulty. To determine the nature of the problem, the primary must send a specific poll. In response to this poll, the secondary returns a Status and Sense message. SOH % R informs the primary that what follows is the Status and Sense message.

SS1 and SS2 is a two-character code which describe the type of problem encountered (Table 11–6). An RVI is never sent by the primary to the secondary.

Example

Assume that the primary has selected a particular device at a station and received a positive acknowledgment.

Message sent by the primary.

Secondary responds with an RVI.

Primary polls the device to determine the nature of the problem.

TABLE 11–6   STATUS AND SENSE CODES[ac]

S S		
S S		
1 2		

S P	A	ESC missing. ETX missing or out of place. Contains invalid buffer address or device address. Example: Specific poll of a device address which that station does not have.
S P	-	Contains invalid BCC sequence.

H	A	Device being copied.
H	S P	Device busy for some other reason.
B	S* P	Device no longer busy.

D	A	Device to be copied not available because the buffer is locked. With some systems, if a protected alphanumeric attribute is in the home position, that buffer cannot be copied. An attempted copy command with the buffer locked will always give a buffer locked status and sense message.
S P	J	Device to be copied not available for some other reason.
B	S* P	Device now available.
B	&[b]	Device not available because it is out of paper or the cover is up.
S P	&	Device not available for other reasons.

[a]Sent only in reply to a specific poll except as noted by an asterisk.
[b]Sent in reply to a general poll as well as a specific poll.
[c](Courtesy of AT&T).

```
S S S       S S        E B P
Y Y O % R T P  D S  A T C A ─────────────
N N H       X A  A P    X C D
```

Since the message from the primary did not start with a command field, the status and sense code (SP A) was sent back.

The primary would now have to correct its error, reselect, and retransmit the original message.

5. Remote Test Request Message (in reply to a poll or a read modified command if R/TST depressed) (Courtesy of AT&T).

S		S	D A A	data in fields	E B P
O	%	/	T	C A A	less nulls	T C A
H		X	1 1 2		X C D
			For each modified field		

The Copy Command

The copy command is sent by the primary to a station controller indicating which device will be copied from. The information to be copied will be sent to a previously selected device. The sole restriction is that the device copied from and the device copied onto must be at the same station. CCC stands for Copy Control Character (Table 11–7). This character identifies what will be copied.

    The sequence of events involved with copying is initiated by the device that will be copied from. If copying from one device onto another is required quite often, a PA key at the device that will be copied from can be programmed to initiate a copy command onto the desired device. Once programmed, depressing that particular PA key will cause a copy only onto the device for which it was programmed. The operator at the device would compose (or have) the message to be copied on the screen and then press the appropriate PA key, because what is on the screen is intended for another device and not the primary station. If a PF key was used, the screen text would be transmitted to the primary. When polled, the station controller would send out a Read Modified

TABLE 11–7 COPY CONTROL CHARACTERS[a]

CCC								
Copy					Print			
Start printer	Sound KD tone	ACs	Protected data	Unprotected data	NL char/line	40 char/line	64 char/line	80 char/line
		×			sp	&	-	0
		×		×	A	J	/	1
		×	×		B	K	S	2
		×	×	×	C	L	T	3
	×	×			D	M	U	4
	×	×		×	E	N	V	5
	×	×	×		F	O	W	6
	×	×	×	×	G	P	X	7
×		×			H	Q	Y	8
×		×		×	I	R	Z	9
×		×	×		[¢	] !	ǀ	:
×		×	×	×	.	$	,	#
×	×	×			<	*	%	@
×	×	×		×	(	)	—	'
×	×	×	×		+	;	>	=
×	×	×	×	×	!	< ¬	?	"

[a](Courtesy of AT&T).

message. The primary would then select the device that will be copied onto. If that device is ready to receive (powered up and not in LOCAL), the station controller returns a positive acknowledgment to the primary. At this time, the primary sends a copy command to the station controller. Up to this time, the operator at the device that will be copied from does not know if the copy request has been sent to the primary; the operator at the device that will be copied onto has no inkling of any copying operation being planned.

On receiving a copy command, the station controller returns a WACK to the primary indicating positive acknowledgment and also requests no transmission from the primary since it will be involved in a copy operation. The station controller then initiates the actual copying with the information going from the device being copied from, to the station controller, to the device being copied onto. While the copying is being carried out, the primary is free to poll or select other stations in the system.

Example

Device 3 at station 5 wants to copy everything on its display onto device 8. The PA1 key has been programmed to inform the primary to send out the appropriate copy command.

```
S S E P S S        E P
Y Y O A Y Y E E " " N A ───────────────►
N N T D N N        Q D
```

Station 5 receives a general poll.

```
S S S      E B P
Y Y T E C % T C A ───────
N N X      X C D
```

E = station 5; C = device 3; % = AID for PA1.

```
S S D   P
Y Y L 1 A ────────
N N E   D
```

Primary acknowledges BCC.

```
S S E P
Y Y O A ────────
N N T D
```

Handshake by secondary.

The primary, by looking at the application program for this device for the % AID, determines what action is required.

```
S S E P S S         E P
Y Y O A Y Y V V H H N A ───────────►
N N T D N N         Q D
```

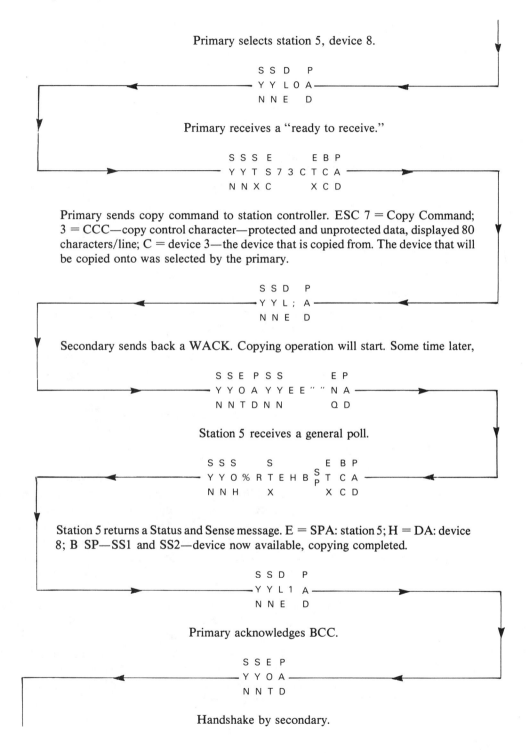

Primary selects station 5, device 8.

```
S S D   P
Y Y L O A
N N E   D
```

Primary receives a "ready to receive."

```
S S S E       E B P
Y Y T S 7 3 C T C A
N N X C       X C D
```

Primary sends copy command to station controller. ESC 7 = Copy Command;
3 = CCC—copy control character—protected and unprotected data, displayed 80
characters/line; C = device 3—the device that is copied from. The device that will
be copied onto was selected by the primary.

```
S S D   P
Y Y L ; A
N N E   D
```

Secondary sends back a WACK. Copying operation will start. Some time later,

```
S S E P S S       E P
Y Y O A Y Y E E " " N A
N N T D N N       Q D
```

Station 5 receives a general poll.

```
S S S     S       E B P
Y Y O % R T E H B S T C A
N N H     X     P X C D
```

Station 5 returns a Status and Sense message. E = SPA: station 5; H = DA: device
8; B SP—SS1 and SS2—device now available, copying completed.

```
S S D   P
Y Y L 1 A
N N E   D
```

Primary acknowledges BCC.

```
S S E P
Y Y O A
N N T D
```

Handshake by secondary.

Device 3 still has the information on its display and the operator has no way of knowing that the copying has been completed.

```
            S S E P S S       E P
          - Y Y O A Y Y V V C C N A -
            N N T D N N       Q D
```

Primary selects device 3 at station 5.

```
            S S D   P
          - Y Y L O A -
            N N E   D
```

Device 3 ready to receive.

```
            S S S E   E B P
          - Y Y T S ? T C A -
            N N X C   X C D
```

ESC ? = Erase all unprotected command. Primary erases all unprotected data from the screen of device 3 as a positive indication that the copying has been completed. The erase-write command with various options of WCC could have also been used.

```
            S S D   P
            Y Y L 1 A -
            N N E   D
```

Secondary returns positive acknowledgment of BCC.

Transparent Text Mode

If Bisync is to be used for the transmission of binary data, floating-point numbers, unique specialized codes, or machine language computer programs, it is possible for these bit patterns to be ASCII codes for certain data link control characters such as SYN, STX, ETB, ENQ, DLE, ETX, or EOT. These bit patterns would have to be *transparent* to the receiver—the receiver would have to know that these bit patterns are part of the text and not intended to be data link control characters. How, then, is the receiver to know when a particular bit pattern is part of the text and when that same bit pattern actually is a control character? The transmitter would have to transmit in transparent text mode. The transmitter enters transparent text mode by transmitting DLE STX instead of just STX. From here on, if a bit pattern is to be a control character, it *must* be prefixed by DLE. To terminate a block, the transmitter must send DLE ETB. When operating in the transparent text mode, the receiver, whenever it receives a DLE, strips it and interprets the next character as a data link control character. If ETX was received without being preceded by DLE, the bit pattern for ETX would simply be treated as part of the data stream and not as a control character.

To transmit DLE as part of the data stream, DLE DLE would have to be transmitted. The first DLE would be stripped.

DLE STX   Initiates transparent mode for the text following.

DLE ETB   Terminates a block of transparent text. The next block could be nontransparent and the message then be terminated by ETX alone.

Example

Assume the primary has already selected the secondary, received a "ready to receive," and is now transmitting the message. Acknowledgment of BCCs is not shown.

First block   Second block   Last block

Three blocks of text are shown, which make up one message. The first and last blocks use transparent text mode. The control characters, labeled "X," contained within transparent text and not preceded by DLE will be treated by the receiver as part of the bit stream of the message and not as control characters.

Examples of Dialogue in a Multipoint Environment

(General poll)

(Handshake)

(General poll of another station)

(PF Form Request key depressed; there was nothing on the device's screen at the time)

```
          S S D   P
          Y Y L 1 A
          N N E   D
```

(Positive acknowledgment of BCC)

```
          S S E P
          Y Y O A
          N N T D
```

(Handshake)

```
          S S E P S S S S       E P
          Y Y O A Y Y S S D D   N A
          N N T D N N A A A A   Q D
```

(Primary selects device that requested the form)

```
          S S D   P
          Y Y L 0 A
          N N E   D
```

(Positive acknowledgment of selection)

```
  S S S E   W D B B               G A       G A         E B P
  Y Y T S 5 C C A A   S C TEXT  S C TEXT  T C A
  N N X C   C 1 1 2                           X C D
```

(Primary sends the form to that device)

```
          S S D   P
          Y Y L 1 A
          N N E   D
```

(Positive acknowledgment of BCC)

```
          S S E P S S S       E P
          Y Y O A Y Y P P " " N A
          N N T D N N A A     Q D
```

(Primary continues with general poll of other stations)

```
          S S E P
          Y Y O A
          N N T D
```

(Handshake)

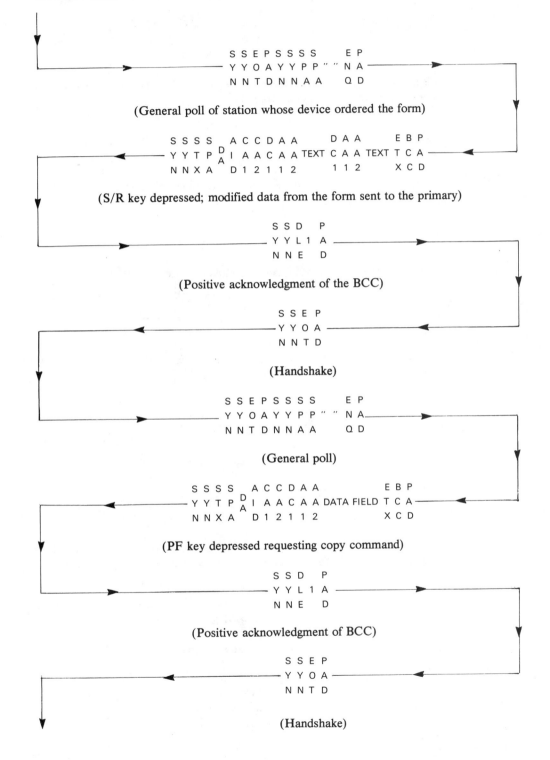

```
              S S E P S S S S       E P
            - Y Y O A Y Y P P " "   N A ------------
              N N T D N N A A       Q D
```

(General poll of station whose device ordered the form)

```
              S S S S    A C C D A A        D A A        E B P
            - Y Y T P D  I A A C A A TEXT C A A TEXT  T C A --
              N N X A A  D 1 2 1 1 2        1 1 2      X C D
```

(S/R key depressed; modified data from the form sent to the primary)

```
              S S D   P
            - Y Y L 1 A ------------------
              N N E   D
```

(Positive acknowledgment of the BCC)

```
              S S E P
            - Y Y O A --
              N N T D
```

(Handshake)

```
              S S E P S S S S       E P
            - Y Y O A Y Y P P " "   N A ------------
              N N T D N N A A       Q D
```

(General poll)

```
              S S S S    A C C D A A                E B P
            - Y Y T P D  I A A C A A DATA FIELD   T C A --
              N N X A A  D 1 2 1 1 2                X C D
```

(PF key depressed requesting copy command)

```
              S S D   P
            - Y Y L 1 A ------------------
              N N E   D
```

(Positive acknowledgment of BCC)

```
              S S E P
            - Y Y O A --
              N N T D
```

(Handshake)

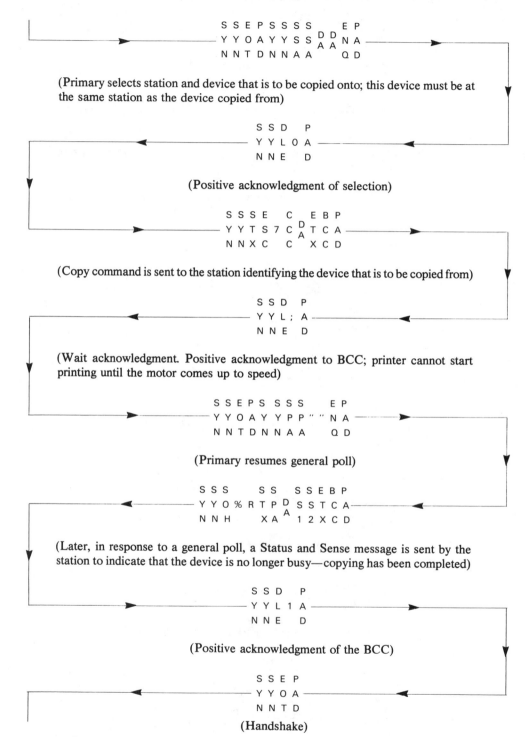

```
                    S S E P S S S       E P
                    Y Y O A Y Y S S D D N A
                    N N T D N N A A A   Q D
```

(Primary selects station and device that is to be copied onto; this device must be at
the same station as the device copied from)

```
                        S S D   P
                        Y Y L 0 A
                        N N E   D
```

(Positive acknowledgment of selection)

```
                    S S S E   C   E B P
                    Y Y T S 7 C D T C A
                    N N X C   C A X C D
```

(Copy command is sent to the station identifying the device that is to be copied from)

```
                        S S D   P
                        Y Y L ; A
                        N N E   D
```

(Wait acknowledgment. Positive acknowledgment to BCC; printer cannot start
printing until the motor comes up to speed)

```
                    S S E P S  S S S      E P
                    Y Y O A Y  Y P P " " N A
                    N N T D N  N A A     Q D
```

(Primary resumes general poll)

```
                  S S S     S S   S S E B P
                  Y Y O % R T P D S S T C A
                  N N H     X A A 1 2 X C D
```

(Later, in response to a general poll, a Status and Sense message is sent by the
station to indicate that the device is no longer busy—copying has been completed)

```
                        S S D   P
                        Y Y L 1 A
                        N N E   D
```

(Positive acknowledgment of the BCC)

```
                        S S E P
                        Y Y O A
                        N N T D
```

(Handshake)

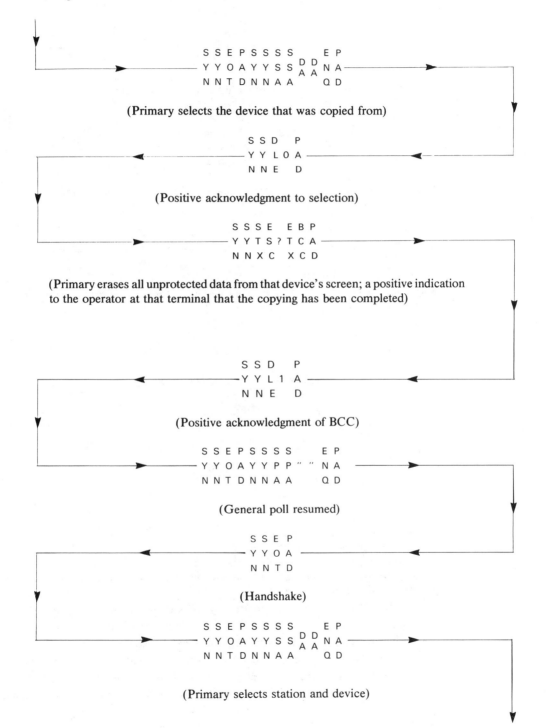

```
              S S E P S S S S       E P
            — Y Y O A Y Y S S  D D  N A —————————————————————————→
              N N T D N N A A  A A  Q D
```

(Primary selects the device that was copied from)

```
                        S S D   P
            ————————————Y Y L 0 A ——————————————————←
                        N N E   D
```

(Positive acknowledgment to selection)

```
                        S S S E   E B P
            ————————————Y Y T S ? T C A ————————————————→
                        N N X C   X C D
```

(Primary erases all unprotected data from that device's screen; a positive indication
to the operator at that terminal that the copying has been completed)

```
                        S S D   P
            ————————————Y Y L 1 A ——————————————————←
                        N N E   D
```

(Positive acknowledgment of BCC)

```
              S S E P S S S S       E P
            — Y Y O A Y Y P P " "   N A ——————————————→
              N N T D N N A A       Q D
```

(General poll resumed)

```
                        S S E P
            ————————————Y Y O A ——————————————————←
                        N N T D
```

(Handshake)

```
              S S E P S S S S       E P
            — Y Y O A Y Y S S  D D  N A ————————————————→
              N N T D N N A A  A A  Q D
```

(Primary selects station and device)

```
        S S D   P
      ← Y Y L < A ─────────────────────────
        N N E   D
```

(RVI: Alarm lamp flashes at controller indicating that it cannot select the device specified by the CPU)

```
        S S E P S S S      E P
      → Y Y O A Y Y P P  D D  N A ─────────→
        N N T D N N A A  A A  Q D
```

(Primary sends a specific poll to extract a Status and Sense message from the station controller)

```
        S S S     S S  S S E B P
      ← Y Y O % R T P  D S S T C A ──────────
        N N H     X A  A 1 2 X C D
```

(Status and Sense message from the controller indicates possible reason for RVI)

```
        S S D   P
      → Y Y L 1 A ─────────────────────────→
        N N E   D
```

(Positive acknowledgment of BCC)

```
        S S E P
      Y Y O A ─────────────────
        N N T D
```

(Handshake)

QUESTIONS

1. Which data link control character(s):
   (a) Immediately precedes each block of text?
   (b) Is used to indicate the end of the message?
   (c) May be used to terminate a block of text which is not the last block of the message?
   (d) Terminates heading information?
2. Which ASCII characters are transmitted for ACK 0 and ACK 1, respectively?
3. What is the response to the first block of text correctly received?
4. What response should be given if the BCC received is bad?
5. WACK is neither a positive nor a negative acknowledgment of a BCC. (T, F)
6. When required RVI can be sent in place of an ACK 0 or an ACK 1. (T, F)
7. In the course of a dialogue, the RVI is sent by the transmitting station to the receiving station. (T, F)
8. When does the primary send an RVI to the secondary?

9. When a station receives an ITB followed by a BCC, it must immediately respond with a ACK 0, ACK 1, WACK, NAK, or an RVI.   (T, F)

10. What response is given to a selection to indicate "ready to receive"?

11. Given the following sequence, draw brackets around all characters that would be involved in the computation of the BCC.

```
S S S                      S                    E B P
Y Y O .................... T ................... T C A
N N H                      X                    X C D
```

12. Given the following sequence:

```
S S S                  I E
Y Y T ................ T T .....................
N N X                  B X
```

   (a) What does ITB indicate?
   (b) What is the purpose of ETX?
   (c) What response is immediately given as a response to the correct reception of this sequence?

13. Given the following sequence:

```
S E     D    G  S        G S S G    E B P
T S 5 9 C K L   Z   S E X          O T C A
                S  P        S P P S
X C     1                           X C D
```

   (a) Where is the cursor after ESC 5 is executed?
   (b) Identify all of the following that have been accomplished by the write control character: (1) start printer; (2) sound KD alarm; (3) KD in LOCAL; (4) ACs reset to unmodified.
   (c) How many characters can you have on one line?
   (d) In which row and column will the S of SEX be stored?
   (e) Can the terminal operator enter a character in the space immediately before SEX?
   (f) How many spaces are allowed the operator to enter the type of sex?
   (g) On this form, the operator could enter information into (*none, 1, 2, 3, more than 3*) locations.
   (h) Assuming that the operator only enters M (for male) properly, since the only visible data transmitted to the CPU are modified data, what portions of the completed form will be sent back?
   (i) In which row and column is the attribute character for the SEX field stored?

14. NULLS and SPACES appear the same way on the display.   (T, F)

15. Which key will the operator have to depress in response to a Read All command?

16. A field in which the operator cannot enter information is called _____ .

17. What is the name for the information that the operator enters?

18. What three characters identify a Status and Sense message?

19. Transparent text is initiated by _____ .

20. If a field is identified as protected, neither the terminal operator nor the transmitter of the form can enter information into that field.   (T, F)

21. A handshake is given by (*the secondary only, the primary only, either the primary or the secondary*).

22. All Status and Sense messages can be sent only in response to a specific poll.   (T, F)

23. In a copy command, the device address supplied is the address of the device that will be copied from.   (T, F)

24. What keys are present on an asynchronous keyboard that are not present on a synchronous keyboard?

25. What problems would be encountered if selections were not specific?

26. What is the difference between a poll and a selection?

PROBLEMS

1. Give the ASCII code in hex for ITB.

2. Write out the entire ASCII *character* sequence for a general poll of station 11.

3. Write out the entire ASCII character response by station 11 if no device at that station had any messages to send and no PF or PA keys had been depressed.

4. Given:

```
S  S  E  P  S  S              E  P
Y  Y  O  A  Y  Y  $  $  J  J  N  A
N  N  T  D  N  N              Q  D
```

(a) The sequence above is a (*general poll, specific poll, selection*).
(b) Which station is involved in this message?
(c) Which device (if any) at this station is involved in this message?

5. Write out the entire ASCII character sequence from the primary to the secondary to send the following form.

```
Row  7    PART (10)
Row 10    QUANTITY (4)
PART and QUANTITY will start in column 10.
```

(1) Clear the screen, sound the KD bell, reset ACs to unmodified, and use 80 characters/line.
(2) The form is protected.
(3) Operator-entered data will be intensified.
(4) Use autoskip and numeric where applicable.
(5) Insert cursor at the first location where the operator can enter data.

6. Write out the entire ASCII character sequence from the primary to the secondary to send the following form:

Row 10   NAME ^S_P (16 spaces) ^S_P S/S ^S_P (11 spaces)   (these 11 spaces include two -'s)

(1) The N of NAME will start in column 7.

(2) Alphanumeric information can be entered into the spaces after NAME.

(3) Only numbers will be entered after S/S.

(4) Command: 64-character line, put KD in LOCAL, reset ACs to unmodified, ring alarm bell.

(5) Use auto skip where applicable.

(6) Operator-entered data: intensified and modified.

(7) Insert cursor for the operator at the first location where operator can enter information.

(8) Everything else on the form should be protected.

12

SYNCHRONOUS DATA LINK CONTROL AND HIGH LEVEL DATA LINK CONTROL

SYNCHRONOUS DATA LINK CONTROL

SDLC is a *bit-oriented protocol*. This means that transmission acknowledgments, polls, and other data link controls are accomplished by a single bit or a group of bits. In contrast, Bisynch is a character-oriented protocol which accomplishes these functions with a data word. With SDLC, text is transmitted in EBCDIC. Bit 7 is always understood to be the least significant bit and bit 0 is the most significant bit. This convention will be adhered to in this text. As shown in Figure 12–1, bit 7 is always transmitted first.

$$b_0 \ldots\ldots\ldots\ldots\ldots b_7 \longrightarrow$$

MSB	LSB
High-order bit	Low-order bit

Figure 12–1   EBCDIC transmission sequence.

SDLC FRAME

Information is transmitted on a SDLC data link in units called *frames*. A frame always starts and ends with a flag. The bit sequence for a flag is $01111110 = 7\text{EH}$. The frame format and the number of bits in each field are shown in Figure 12–2.

There are several variations of how a terminating flag of one frame and the starting flag of the next frame may be transmitted. These variations are shown in Figure 12–3.

210

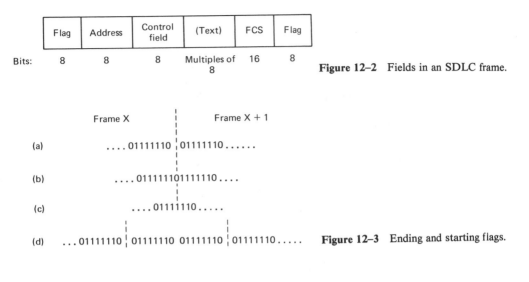

Bits: 8   8   8   Multiples of   16   8
                                          8

Figure 12–2 Fields in an SDLC frame.

Frame X   Frame X + 1

(a)   .... 01111110 ¦01111110 ......

(b)   .... 0111111101111110 ....

(c)   .... 01111110 .....

(d)   ... 01111110 ¦ 01111110 01111110 ¦ 01111110 .....

Figure 12–3 Ending and starting flags.

(a) Two totally separate flags.
(b) The last 0 of the terminating flag is also the first 0 of the starting flag.
(c) The terminating flag of one frame is also the starting flag of the next frame.
(d) Additional flags inserted between the terminating flag of one frame and the starting flag of the next frame.

The transmitted address is always the address of the secondary. If the primary is transmitting, the address would mean: "This message is for...." If the secondary is transmitting, the address would mean: "This message is from...." The primary does not have an address. Secondaries may have individual, group, and/or broadcast addresses. A *group address* is an address common to two or more secondaries in the data link but not to all. A *broadcast address* is common to all secondaries. FFH is used as the broadcast address. If the primary sent information addressed to FFH, all secondary stations in the data link would receive the information. The address 00H is not assigned to any station. This address is called a *void address* and may be used by the primary to troubleshoot the system.

The *frame check sequence* (FCS) is computed using CRC-16 (Chapter 4). The CRC is computed on the bit stream following the starting flag to and including the text if present (Figure 12–4). CRC-16 generates 16 bits for error detection.

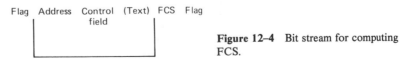

Flag Address Control (Text) FCS Flag
                field

Figure 12–4 Bit stream for computing FCS.

The control field has three formats: information, supervisory, and unnumbered (unsequenced). The type of format used is determined by the two least significant bits of this field:

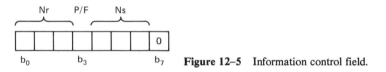

$$
\begin{array}{cccc}
b_6 & b_7 \\
X & 0 & \text{Information} & \text{(text always present)} \\
0 & 1 & \text{Supervisory} & \text{(text never present)} \\
1 & 1 & \text{Unnumbered} & \text{(text may be present)}
\end{array}
$$

Information Format

Figure 12–5 shows the bit pattern of an information control field. Bit 3 is the *Poll/Final* (P/F) *bit.* If the transmission is from the primary to the secondary, this bit is a poll bit. If $P = 0$, the frame is not a poll and the secondary is not invited to answer. If $P = 1$, the frame is a polling frame and the secondary is required to answer after it receives this frame. If the transmission is from the secondary to the primary, this bit is a final bit. If $F = 0$, the frame sent by the secondary is not the last frame and the primary will not respond. If $F = 1$, the frame sent is the last frame that the secondary wishes to transmit in this sequence and the primary will respond appropriately.

Nr　　P/F　　Ns

| | | | | | | | 0 |

b_0　　　　　b_3　　　　　b_7　　**Figure 12–5**　Information control field.

The Nr (number received) transmitted by the primary is related to the Ns (number sent) transmitted by the secondary; the Nr transmitted by the secondary is related to the Ns transmitted by the primary. If information format is used, text is sent in numbered frames. The sender identifies which frame is being sent by sending Ns, a 3-bit binary number. If $Ns = 010$, frame 2 is being sent. Whoever receives the information transmits the Nr. Nr identifies the number of the next frame expected. If the secondary transmitted an $Ns = 2$ and received an $Nr = 3$, the primary is acknowledging successful reception through frames $Nr - 1$ and is expecting to receive frame 3 next.

Since each frame has an FCS, does each frame have to be acknowledged immediately? No. A maximum of seven frames can be transmitted without requesting confirmation. Confirmation is requested by making the P/F bit a 1. Acknowledgment (Nr) can be made either with an information frame if the receiving end has information to transmit or with a supervisory frame if there is no text to send.

Primary　　　　　　　　　　　　Secondary

Nr　　　　Ns　　　　　　　　　　Nr　　　　Ns

Supervisory Format

The supervisory control field (Figure 12–6) is identified by 01 in its two least significant bits.

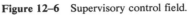

Figure 12–6 Supervisory control field.

The secondary clears a RNR by sending an information frame with F = 1 or a RR or REJ frame with F = 0 or 1. The primary clears a RNR by sending an information frame with P = 1 or a RR or REJ frame with P = 0 or 1.

Example

Assume that the primary is transmitting to the secondary and all Nr and Ns counters are initially zeroed.

Primary			Secondary			
Nr	P	Ns	Nr	F	Ns	
000	0	000 0				Primary transmits three information frames.
000	0	001 0				The third frame is a poll.
000	1	010 0				
			011	0	000 0	Secondary acknowledges with Nr = 3.
			011	1	001 0	It sends two information frames back.
010	0	011 0				Primary acknowledges errorless reception of two frames with
010	0	100 0				Nr = 2. Send six information frames.
010	0	101 0				
010	0	110 0				
010	0	111 0				
010	1	000 0				
			110	0	010 0	Since Nr = 6, the secondary acknowledges all frames
			110	0	011 0	through Nr − 1 = 5 and requests retransmission of
			110	1	100 0	frame 6. Since the primary does not know whether
						only frame 6 had an FCS error or whether any subsequent
						frames also had an error, the primary must retransmit
						everything from the sixth frame on.
101	0	110 0				Primary acknowledges secondary's frame 4 with Nr = 5.
101	0	111 0				Continues to transmit its text starting with frame 6.
101	0	000 0				Note the counter rollover.
101	1	001 0				
			010	1	00 01	Secondary confirms all frames with Nr = 2.
						Since the secondary has no text to send, supervisory
						format is used.
101	0	010 0				Primary sends five additional frames.
101	0	011 0				
101	0	100 0				
101	0	101 0				
101	1	110 0				
			101	1	00 01	Secondary acknowledges through frame 4 and requests
				(RR)		retransmission of frame 5 and what follows. Since the Nr
				or		indicates this, note the redundancy produced by bits 4 and 5
			101	1	10 01	of the reject format.
				(REJ)		

If more than seven frames are sent before requesting acknowledgment, ambiguity can be introduced.

Transmitted Ns	
000	If this many frames were transmitted before requesting
001	acknowledgment, and an $Nr = 0$ was returned, there would be
010	confusion as to whether the first frame had an FCS error
011	and everything had to be retransmitted or if everything was
100	received without error and the next expected frame number was 000.
101	
110	
111	

Unnumbered Format

If the two least significant bits of the control field are both 1's, an unnumbered format is being used (Table 12–1). As the name implies, frames sent with the control field in this format are not numbered. The control field contains neither Nr nor Ns. If Ns is not present, these frames cannot be acknowledged with an Nr. Similarly, since these control fields do not contain an Nr, this type of format cannot be used to acknowledge numbered frames (frames containing an Ns). Only those control characters used in a multipoint data link will be discussed.

TABLE 12–1

Binary configuration						
b_0		b_7	Acronym[a]	Sent by the primary	Sent by the secondary	Text permitted
000	P/F	0011	UI	×	×	×
000	P	0111	SIM	×		
000	F	0111	RIM		×	
100	P	0011	SNRM	×		
000	F	1111	DM		×	
010	F	0011	RD		×	
010	P	0011	DISC	×		
011	P/F	0011	UA	×	×	
100	F	0111	FRMR		×	×

[a]Obsolete acronyms and their meaning are shown in parentheses:

UI	Unnumbered information (NSI—Nonsequenced Information)
SIM	Set Initialization Mode
RIM	Request Initialization (RQI—Request Initialization)
SNRM	Set Normal Response Mode
DM	Disconnect Mode (ROL—Request On-Line)
RD	Request Disconnect (RQD—Request Disconnect)
DISC	Disconnect
UA	Unnumbered Acknowledgment (NSA—Nonsequenced Acknowledgment)
FRMR	Frame Reject (CMDR—Command Reject)

RIM is transmitted by the secondary to notify the primary of the need for a SIM command. A SIM command sent to the secondary would initiate system-specified procedures for the purpose of initializing link-level functions. The primary and secondary Nr and Ns are reset by this command. The secondary is expected to answer with UA. After initialization, the secondary is either in the normal response mode or the disconnect mode. If it is in the *disconnect mode,* the secondary can transmit and receive only unnumbered commands. In this mode, the secondary is effectively off-line. In the *normal response mode,* the secondary is capable of transmitting and receiving information, supervisory, and unnumbered frames. The secondary cannot go into the normal response mode or the disconnect mode on its own but must be placed in that mode by the primary. Assume initially that the secondary is already in the disconnect mode. Assume also that the primary sends an information frame to the secondary. The secondary does receive it but cannot react to it. The secondary would reply with the unnumbered frame of DM. In effect, it is telling the primary: "I am in the disconnect mode and I cannot receive information frames." The primary would then transmit a SNRM frame which would place the secondary in the normal response mode. When the primary receives a UA from the secondary, it could then retransmit the information frame(s). If the secondary wished to return to the disconnect mode after receiving the primary's message, it would return a supervisory frame to provide the primary with a numbered acknowledgment with the F bit a 0. The following frame would be an unnumbered frame—RD. The primary would respond with DISC, which would return the secondary to the disconnect mode. The secondary would be expected to return a UA. SNRM also resets Nr and Ns of the primary and the secondary.

To transmit FRMR, the secondary must be in the normal response mode. The FRMR indicates to the primary that the previously received frame by the secondary is invalid. Reasons why a frame may be invalid:

1. The control field cannot be implemented by the secondary.

Example

Secondary receives a control field of

011 P 11 11   (LSB on right)

The two LSBs identify this as an unnumbered frame, but there is no command that corresponds to the remaining bits.

2. The information field is too long to fit into the receiving station's buffers.
3. The received control field does not allow an information field in that frame.

Example

The secondary receives a supervisory control field. Since no text is permitted with this field, the secondary expects the next two bytes to be the FCS followed by a flag. If a flag is not detected until after 20 bytes have been received following the control field, it presumes that a text field has been included.

4. The Nr received from the primary is invalid.

Example

Secondary transmits three numbered frames: Ns = 000, 001, and 010. In response, it receives an Nr = 011, implying that all frames were received correctly. The secondary sends four more frames: Ns = 011, 100, 101, and 110. In reply, it now receives an Nr of 001. Only an Nr of 011, 100, 101, 110, or 111 would make any sense. The secondary sends back a FRMR.

The text field allowed with FRMR is fixed and is used to identify the reason why the received frame was invalid. The format of a Frame Reject frame is shown in Figure 12–7.

The two bracketed bytes are shown with the order of the arrow to the left (leftmost bit is transmitted first). Ns and Nr are the current values of these two counters at the secondary.

w = 1   if the secondary received a command which is invalid or if it cannot be implemented
x = 1   if a prohibited information field is received
y = 1   if there was a buffer overrun (information field too long)
z = 1   if the received Nr disagrees with the transmitted Ns

The four 0's following z are simply stuffed to complete this byte.

Once the secondary transmits a FRMR, it cannot release itself from this condition, nor will it act on the frame that caused this condition. It will repeat the FRMR whenever it responds except to an acceptable mode-setting command such as DISC, SIM, or SNRM which resets the FRMR condition. Following is a sample dialogue in a multipoint, full-full-duplex system (Figure 12–8). Station A is initially in NRM and station B is initially in DM.

Flag	Address field	FRMR control field	Control field of rejected frame	Ns	Nr		FCS	Flag
				0 xxx 0 xxx	w x y z 0000			

Figure 12–7   Frame Reject frame format.

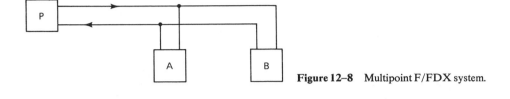

Figure 12–8 Multipoint F/FDX system.

Flag A RR FCS Flag
7E C1 11   7E

Primary polls Station A

    Flag A Info (Text) FCS Flag
    7E C1 00   7E

Station A sends first information frame;
Nr = Ns = 0, F = 0

    Flag A Info (Text) FCS Flag
    7E C1 02   7E

Station A sends second information
frame; Nr = F = 0, Ns = 1

    Flag A Info (Text) FCS Flag
    7E C1 14   7E

Station A sends third information frame;
Nr = 0, F = 1, Ns = 2

Flag A Info (Text) FCS Flag
7E C1 60   7E

Primary sends text reply; Nr = 3,
P = 0, Ns = 0

Flag A Info (Text) FCS Flag
7E C1 72   7E

Primary sends second information frame;
Nr = 3, P = 1, Ns = 1

    Flag A REJ FCS Flag
    7E C1 39   7E

Station A had an FCS error with second
frame; Nr = 1, F = 1

Flag A Info (Text) FCS Flag
7E C1 72   7E

Primary retransmits second frame;
Nr = 3, P = 1, Ns = 1

    Flag A RR FCS Flag
    7E C1 51   7E

Station A acknowledges the second
frame; Nr = 2, F = 1

Flag B Info (Text) FCS Flag
7E C2 10   7E

Primary sends information frame to
Station B; Nr = Ns = 0, P = 1

    Flag B DM FCS Flag
    7E C2 1F   7E

Station B indicates that it is in DM

Flag B SNRM FCS Flag
7E C2 93   7E

Primary places station B in NRM

    Flag B UA FCS Flag
    7E C2 73   7E

Station B returns an unnumbered
acknowledgement

Flag B Info (Text) FCS Flag
7E C2 00   7E

Primary sends station B the first
information frame; Ns = Nr = 0, P = 0

Flag B Info (Text) FCS Flag
7E C2 02   7E

Primary send station B the second
information frame; Ns = 1, P = 0,
Nr = 0

Flag B Info (Text) FCS Flag
7E C2 04   7E

Primary sends station B the third
information frame; Ns = 2, P = 0,
Nr = 0

Flag B Info (Text) FCS Flag
7E C2 16   7E

Primary sends station B the fourth
information frame; Ns = 3, P = 1,
Nr = 0

Flag B RR FCS Flag 7E C2 91 7E		Station B acknowledges all frames; Nr = 4, F = 1
Flag B DISC (Text) FCS Flag 7E C2 53 7E		Primary sends a disconnect command to station B; erroneously sends text along in the same frame
Flag B FRMR DISC —— —— FCS Flag 7E C2 97 53 80 02 7E		Station B replies with a frame reject; DISC is the control field of the rejected command 80 Nr = 4, Ns = 0: current counter of station B 02: x = 1: text not allowed
Flag B DISC FCS Flag 7E C2 53 7E		Primary sends disconnect command to station B
Flag B UA FCS Flag 7E C2 73 7E		Station B replies with an unnumbered acknowledgment

Zero Stuffing (Zero-Bit Insertion)

The flag bit pattern may occur in a frame where this pattern is not intended to be interpreted as a flag.

Examples

1. The transmission of an "=" as part of the text. The EBCDIC for "=" is 01111110.
2. A control field using information format where Nr = 3, P/F = 1, and Ns = 7. This control field would also yield a bit pattern of 01111110.

Anytime this pattern is received, it is interpreted as a flag. Therefore, the receiver would interpret the frame to be ending somewhere in the middle of the actual message. To eliminate this ambiguity, a zero is inserted automatically after five consecutive ones found between the starting and ending flags of a frame. This process is clearer if the sequence of events is understood:

1. Message is composed:

Control (Text)
Field

2. The station controller inserts the station address:

Address Control (Text)
Field

3. The FCS is computed:

Address Control (Text) FCS
Field

4. Zeros are now stuffed after any five contiguous ones found anywhere from the address to the FCS inclusive.

5. Starting and ending flags are now added before transmission.

The stuffed 0's are not included in the computation of the FCS since they are added after the FCS has been computed. In fact, the FCS may have 0's stuffed into it. This process prevents the occurrence of a flag pattern anywhere between the starting and ending flags.

At the receive terminal, the process is reversed. Once the starting flag is identified, any zero following five successive ones is stripped. Six successive ones followed by a zero will be interpreted as the terminating flag.

Example

Station B transmits "C?" as the information field.

"o" is used to indicate a stuffed zero.

Order of the arrow is to the left.

01111110 01000011 011111o10 11111o110 11000011 111o10110

Flag   Add. = B   Cont. Fld   " = 7F   C = C3   ? = 6F

11111o110 FCS 01111110 111111111111...

" = 7F   Flag   Idle Line 1's

Message abort. Anytime after the starting flag is received, the reception of *seven to 14 consecutive 1's* indicates an abort condition. Zeros are not stuffed into an abort signal. An abort simply tells the receiving station to disregard everything sent thus far in the current frame. Clock synchronization is still maintained by the receiver during an abort condition. An abort condition is used to terminate a frame and immediately begin another frame of higher priority.

Idle line state. If the data link is operational but there is no SDLC control or information transmission currently in progress, the idle line state exists. A station perceives the existence of an idle link when it receives *15 or more consecutive 1's.*

Comparison of Bisynch and SDLC

Bisynch is a character-oriented protocol, whereas SDLC is a bit-oriented protocol. ASCII or EBCDIC can be used with Bisynch, whereas EBCDIC is always used with SDLC. Both Bisynch and SDLC are synchronous protocols. For error detection, ASCII uses a block check sequence derived from Longitudinal Redundancy Checking (LRC) which consists of one byte; EBCDIC uses a block or frame check sequence derived from Cyclic Redundancy Checking (CRC) which consists of two bytes. When operating in a multipoint environment, approximately 256 bytes are transmitted per frame/block for both protocols. Since Bisynch is a character-oriented protocol, it has a greater difficulty handling the occurrence of data bytes that have the bit patterns of control characters within the text. Bisynch must resort to the transparent text mode. SDLC does not have similar problems. The occurrence of flag patterns within a frame is the only thing that would present a misunderstanding, and this is remedied by the process of zero bit stuffing.

Invert-on-Zero Coding

For the receiver to maintain bit synchronization, there must be frequent variations in the levels of the received signal—no prolonged receptions of constant 1's or constant 0's. Zero bit stuffing eliminates the possibility of receiving more than six consecutive 1's as part of any unaborted frame. The possibility of receiving a long string of 0's as part of a frame still exists. Data terminal may possess the option of non-return-to-zero inverted (NRZI) encoding of the data bits of a frame before these bits are applied to the RS 232C interface. With this form of encoding, there is no change in the signal level to transmit a 1; to transmit a 0, the signal level is always inverted from what it was. An example of NRZI encoding is shown in Figure 12–9. The signal level is assumed high before the reception of the first one of the actual bit stream.

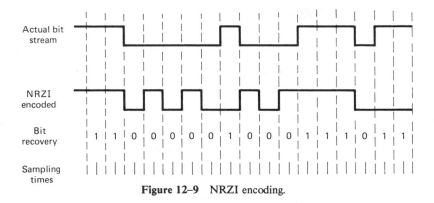

Figure 12–9 NRZI encoding.

HIGH LEVEL DATA LINK CONTROL

In 1975, the International Standards Organization (ISO) defined a set of standards known as High Level Data Link Control. HDLC is a superset of IBM's SDLC. HDLC is divided into smaller subdivisions that, when combined, outline the frame structure, control standards, and class of operation of a bit-oriented Data Link Control (DLC). Since HDLC is a superset of SDLC, only the added capabilities as prescribed by the indicated ISO standards will be described.

ISO 3309-1976(E).   This standard defines the frame structure, delimiting sequence, and transparency mechanism employed by HDLC. These are all identical to SDLC except that HDLC has extended address capability and the FCS is checked differently. In HDLC, the address field may be extended recursively. A logic 1 for b_0 identifies that byte as the last byte of the address field: a logic 0 indicates that the byte immediately following is part of an extended address. (*Note:* The ISO standard identifies the low-order bit as b_0. In SDLC, the low order bit is designated b_7.) Basically, 7 bits are available for address encoding within each address byte. A given system cannot use both basic and extended addressing. Figure 12–10 shows an example of an HDLC frame using extended addressing with three address bytes.

01111110   0XXXXXXX 0XXXXXXX 1XXXXXXX   01111110

Beginning   Three Byte Address Field   Ending
Flag   Flag

Figure 12–10   HDLC frame using extended addressing.

HDLC uses a CRC, generated by a polynomial specified by CCITT V.41, as the FCS. The transmitting station computes a 16-bit CRC and transmits it just prior to the ending flag. The receive station computes the CRC in the same manner as SDLC except that the computation includes the received CRC. If the transmission is without error, the CRC computed at the receiver is always F0BBH.

ISO 4335-1979(E).   This standard defines the elements of procedures for HDLC. The control field, the information format, and the supervisory format have increased capability over SDLC. HDLC also has two additional operational modes.

Operational modes.   The three operational modes of SDLC—initialization mode, normal response mode, and disconnect mode—are described in SDLC under unnumbered formats. HDLC has the two following additional operational modes:

1. *Asynchronous Response Mode (ARM):*   In this mode, secondary stations are allowed to send unsolicited response frames. For a secondary station to transmit, it need not have received a frame with the P bit set. However, if a secondary receives a command with the P bit set, it must reply with a frame with the F bit set. Additional frames may be transmitted after the response to poll has been sent. Final indicates only a response to a poll and is not an indication of the last frame of a message.
2. *Asynchronous Disconnect Mode (ADM):*   ADM is identical to NDM except that a secondary may initiate a DM or RIM at any time.

Control field.   HDLC has an extended mode capability. In the extended mode, the control field is increased to 16 bits. Seven bits are assigned to the Ns and 7 bits are assigned to the Nr. In the extended mode, there may be as many as 127 unacknowledged frames outstanding at any one time.

Information format.   EBCDIC is the only code used with SDLC. The information field, prior to zero substitution, must have a multiple of 8-bit characters. HDLC is not byte oriented: a character or word within the information field of an HDLC frame may have any number of bits as long as all of the character and word lengths are the same. Consequently, the coding used with HDLC is less restricted than with SDLC.

Supervisory format.   The supervisory format for HDLC, like SDLC, includes ready to receive (RR), ready not to receive (RNR), and reject (REJ). However, HDLC includes, in addition, a selective reject (SREJ). A selective reject calls for retransmission of only the frame identified by Nr. Only one SREJ may be established by a station at a time. Additional SREJs may be established after the first SREJ has been cleared by receipt of an I-frame with an Ns equal to the SREJ's Nr.

ISO 6159-1980(E) (unbalanced) and ISO 6256-1981(E) (balanced). ISO is currently merging these two standards into ISO 7809. These two standards identify the class of operations as either balanced or unbalanced. The defined class of operation identifies the procedures necessary to establish the link-level protocol.

Unbalanced operation. Circuit configuration consists of a primary station and one or more secondary stations. The data link operation may be half-duplex or full-duplex. Channel access is through polling using the NRM.

Balanced operation. Circuit configuration consists of only two stations. Each station can set up or disconnect the link. Channel access is through contention using ARM. Data link operation may be half-duplex or full-duplex.

QUESTIONS

1. SDLC is a _____ -oriented protocol.
2. Identify the code used with SDLC.
3. Which bit is the low-order bit (bit position transmitted first) in EBCDIC?
4. What are the hex characters for a flag?
5. Excluding flags, identify the four fields that comprise an SDLC frame.
6. In SDLC, the transmitted address is always the address of what station?
7. An address common to all secondary stations is called a _____ address. What is the hex sequence for this address?
8. An address common to more than one but not all of the secondary stations is called a _____ address.
9. Identify the hex address that is never assigned to a secondary station.
10. Identify the three control field formats that are used with SDLC.
11. Which of the formats in Question 10 allows the transmission of text information?
12. The P/F bit is included with all three of the control field formats.   (T, F)
13. All frames containing information are numbered.   (T, F)
14. Which control field format is used to acknowledge the reception of numbered frames, but does not allow the transmission of an information field?
15. What is the purpose of the Nr bits?
16. What is the purpose of the Ns bits?
17. An information frame is identified by making bit position _____ of the _____ field a logic _____ .
18. What is the maximum number of information frames that can be sent before an acknowledgment is required with SDLC?

19. What are the low-order bits for unnumbered control characters?
20. What follows the control character in a supervisory frame?
21. What is the secondary's response to an invalid command from the primary?
22. What is the secondary's response to a SNRM command from the primary?
23. Supervisory control characters cannot be used to acknowledge previously received numbered frames. (T, F)
24. To distinguish a flag from an EBCDIC = sign, a process of _____ is used.
25. The frame check sequence is made up of 8 bits. (T, F)
26. Which unnumbered commands reset the Nr and the Ns counters to all 0's?
27. Which response indicates that the secondary station is in the disconnect mode?
28. When in the disconnect mode, the secondary cannot transmit either of the three frame formats. (T, F)
29. What is the purpose of NRZI encoding?
30. Flags are zero-inserted. (T, F)
31. What is the maximum number of outstanding (unacknowledged) frames allowed with HDLC in the basic control mode? In the extended control mode?
32. What is the delimiting sequence used with HDLC?
33. What supervisory condition is included with HDLC that is not included in SDLC?
34. EBCDIC is the only code allowed in HDLC. (T, F)
35. How does HDLC achieve transparency?
36. Which bit in the address field of an HDLC frame indicates that additional address bytes follow when in the extended addressing mode?

PROBLEMS

1. 11111110111111011111101110000000111110101100111010110011011111100
01001111100101100101011000110011101111110111111111111111111111111
The message above is an SDLC data stream; the order of the arrow is to the left.
   (a) What is the format of the frame above?
   (b) What is the hex address of the secondary involved in this transmission?
   (c) If an Ns is associated with this frame, what is it?
   (d) If an Nr is associated with this frame, what is it?
   (e) This frame is the last of a series of frames, or the only frame in this message. (T, F)
   (f) Excluding flags, how many characters are in this frame?
2. Stuff 0's where required. Order of the arrow is to the left.

1 0 0 1 1 1 1 1 1 1 1 1 1 0 0 1 1 1 1 0 1 0

3. SDLC message in EBCDIC:

   (a) What is the address of the secondary?

(b) The message is from a primary to a secondary station.   (T, F)

(c) How many bytes follow p in the message above?

4. If the primary sent four successive frames starting with $Ns = 2$, the secondary would return a positive acknowledgment of all four frames by returning an Nr of_____ .

5. Complete the following chart for the control characters given.

Control Character	Format	P/F, $\overline{P}/\overline{F}$	Ns	Nr	Command/Response
$A6_{Hex}$					
N_{EBCDIC}					
53_{Hex}					
68_{Hex}					
22_{Hex}					

Enter P/F if this bit is a 1, but the transmission direction is not determined. Enter P or F if the transmission direction is known.

6. Determine the hex characters for the control field that are required to convey the following information:

(a) Represent an information frame sending frame 3. It is the final frame and frame 5 is the next frame expected to be received.

(b) Represent the primary sending a supervisory frame that is a poll. It is ready to receive and frame 6 is the next frame expected to be received.

(c) Represent the secondary sending an unnumbered frame telling the primary that it is in the disconnect mode.

(d) Represent the secondary sending information frame 7. It is a final frame and the next frame it expects to receive is frame 3.

7. Draw the level diagram that the following bit stream, when encoded in NRZI, would produce.

13

PACKET SWITCHING
AND LOCAL AREA NETWORKS

SWITCHING TECHNIQUES

Packet switching is the transmittal of small bundles of information through switches to the intended destination. This sounds logical. However, if you did not know what packet switching was before, probably very little was added to your knowledge of this subject. Let's take a step back and see how it came about and what the alternatives are. The rapid growth in the number of personal computers, small business computers, mainframe computers, and terminals has created a situation where more and more people have the need to exchange digital information with each other. One method of accomplishing this is by making direct connections between all of the terminals. If n terminals are to be interconnected, the first terminal would require $n-1$ connections; the second would require $n-2$ additional connections; the third would require $n-3$ additional connections; and so on. These connections are shown in Figure 13–1. Mathematically, the sum of a series of integers $1, 2, \ldots, k$ is computed from the formula $k(k+1)/2$. Therefore, to interconnect all of our terminals, $(n-1)n/2$ connections are required. If only a few devices are all connected to each other, the number of connections is not too high. However, as the nth terminal is added to the network, an additional $n-1$ connections are required. Soon, this number of dedicated connections becomes totally impractical. The alternative is to use a switch to which all of the terminals are connected. If there are 100 terminals, then 100 connections would be required to connect each terminal to the switch. The switch would have the capability of linking any two terminals that are connected to it. A switched network is shown in Figure 13–2. For applications where a large number of terminals are involved, the cost of the switch is offset by the savings realized in the reduction of the number of connections. As a network becomes larger, the inclusion of more than one switch may be required. If so, interswitch trunks would be needed.

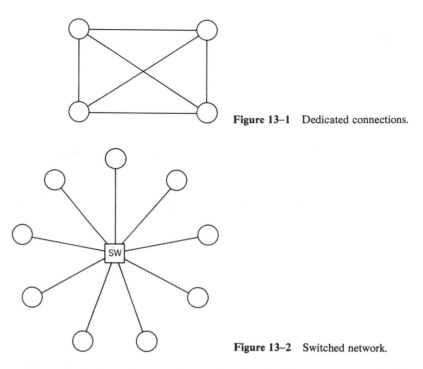

Figure 13-1 Dedicated connections.

Figure 13-2 Switched network.

There are three basic switching techniques: circuit switching, message switching, and packet switching.

Circuit Switching

The public telephone network in the United States uses circuit-switching principles. Once a call has been established, the network provides a dedicated line for the duration of the call. The time required to establish the connection is called the *setup time,* which may be quite long. Since one line is allocated for each call, blocking may occur. *Blocking* occurs when the user is denied a connection through the network because no facilities are available between the points of origin and destination. If circuit switching is used for the transmission of digital information, the terminals at both ends of the connection must be compatible—they must use a common protocol, bit rate, and character set. The equipment used to interface the data terminal equipment to the network is voice compatible and common calling procedures may be used. Once a call is established, information is transferred in real time. Inefficiencies in this type of switching arise after a line has been assigned. If the user does not use the full capacity of the line for the entire time, no other user may have access to it during this period.

Message Switching

Message switching is a form of *store-and-forward* network. Each switch within the network has a message storing capacity. The user inputs information into the network and the network transfers the information to the next switch or to the final destination when it is convenient to do so. Consequently, the delay time from message transmission to reception varies and may be quite long. Blocking cannot occur. Once the information has entered the network, it is converted to a format more suitable for transmission within the network. At the terminating end, the information is converted to a format compatible with the receiving data terminal equipment. Therefore, the source and destination terminals need not be compatible. The basic unit of transmission is still the entire message. Message switching is more efficient than circuit switching since messages received at a time of peak load can be delayed and transmitted when the load has decreased. This efficiency is achieved at the cost of not transmitting messages in real time.

Packet Switching

With packet switching, the user's data are divided into small segments called *packets*. Since a packet may be held in memory for a short period of time, packet switching is called a *hold-and-forward* network. The individual packets of a long message need not take the same path to the destination. Since the packets are small, the hold time is also small and the information is transmitted in near real time. The source and destination equipment need not be compatible since the network makes the necessary adjustments. Blocking cannot occur. The price paid for this high efficiency and near real-time transmission is the requirement of complex switching arrangements and complex protocols.

A circuit switch is called a *transparent switch*. It does nothing but link the transmitted information to the intended destination. It is totally transparent to the user. A message switch or a packet switch is called a *transactional switch* because it does more than connect the send station to the receiver. It can either store the data temporarily or change the data format. At the receive end, the data may again be changed so that they are compatible with the user's terminal equipment.

THE INTERNATIONAL STANDARDS ORGANIZATION (ISO) PROTOCOL HIERARCHY

In order to facilitate the intercommunication of data-processing equipment, the ISO developed a standard protocol hierarchy (Figure 13–3). The intent of this layered structure of network protocols is to allow a network of open systems to be logically composed of smaller subdivisions called levels or layers. Each layer is further divided into smaller subsystems corresponding to the intersection of the system layers. The basic idea of layering is that each layer adds value to services provided by the set of lower layers in such a way that the highest layer is offered the full set of services needed to run distributed applications. Layering thus divides the total system problem into smaller pieces.

Physical layer.   The physical layer is the lowest and most fundamental layer of the ISO hierarchy. This layer specifies the physical, electrical, functional, and

Level 7	Application
Level 6	Presentation
Level 5	Session
Level 4	Transport
Level 3	Network
Level 2	Data link
Level 1	Physical

Figure 13–3 ISO protocol hierarchy.

procedural standards for accessing the network. Several documents are being established to further define the physical layer. The functions specified in the physical layer are similar to those outlined in the EIA RS 232C interface (Chapter 6).

Data link layer.   The primary function of the data link layer is to ensure the reliable transfer of user data over the data link. This layer is responsible for the communication between nodes within the network. The data link layer is the final framing of the message envelope. It facilitates the orderly transmission of data between nodes and allows for error detection and retransmission.

Network layer.   The network layer determines the most appropriate network configuration for the function provided by the network. Examples of network configurations are dial-up, leased, or packet. Flow control procedures are initiated at this level and switched services are initiated through a data call establishment procedure.

Transport layer.   The transport layer acts as an interface between the session and network layers. The transport layer is concerned with economic efficiency. One important function performed by this layer is controlling end-to-end data integrity, which includes routing, message segmenting, and error recovery.

Session layer.   The session layer is responsible for availability. A session is a temporary relationship. Activating a session makes available the appropriate resources such as buffer storage and processor capacity. Session control includes such things as network log-on and user authentication. The session layer coordinates interaction between end application processes.

Presentation layer.   The presentation layer defines protocols that communicate syntax (or representation). The presentation layer functions include formatting, encoding, encrypting/decrypting of messages, dialogue management, synchronization, interruption, and termination.

Application layer.   The application layer is analogous to the general manager of the network. It is responsible for the transfer of information between application processes and the end-to-end encryption. The application layer controls the sequence of events between the computer application and the user or another application. It also controls the sequence of activities within the application itself. The application layer is the layer that communicates directly with the user's application program or process.

CCITT X.1: INTERNATIONAL USER CLASS OF SERVICE IN PUBLIC DATA NETWORKS

The CCITT X.1 standard divides 11 classes of service into three basic modes of operation: start/stop (S/S), synchronous (SY), and packet (PM). Data entering a public data network must be in one of these three forms.

Start/stop mode (classes 1 and 2). In the start/stop mode, data are transferred from the user to the network in an asynchronous data format (each character is framed between a start and 1 or 2 stop bits). IBM's 83B protocol and the Bell System's 8A1/8B1 selective calling arrangement (Chapter 10) are examples of common asynchronous protocols. Call control signaling is done in International Alphabet #5 (ASCII), and data transfer signaling is in accordance with CCITT X.4 with the following bit/second and bit/character structure:

Class 1:   300 bps, 11 bits/character
Class 2:   50 bps, 7.5 bits/character
           100 bps, 7.5 bits/character
           110 bps, 11 bits/character
           134.5 bps, 9 bits/character
           200 bps, 11 bits/character

Synchronous mode (classes 3 thru 7). In the synchronous mode, data are transferred from the user to the network in a synchronous data format (each transmission is preceded with a unique 8-bit synchronizing character). IBM's 3270 BSC (Chapter 11), Burrough's BASIC, and UNIVAC's UNISCOPE are examples of common synchronous protocols. Call control and data transmission rates are identical to private-line applications. The classes of service for synchronous modes are:

Class 3:   600 bps
Class 4:   2.4 kbps
Class 5:   4.8 kbps
Class 6:   9.6 kbps
Class 7:   48 kbps

Packet mode (classes 8 through 11). In the packet mode, data are transferred from the user to the network in a frame format. Each frame is delimited by a beginning and an ending flag. The ISO HDLC frame format (Chapter 12) is a common data link control procedure. Within the frame, call control and data transfer are in accordance with CCITT X.25 at the following transmission rates:

Class 8:   2.4 kbps
Class 9:   4.8 kbps
Class 10:   9.6 kbps
Class 11:   48 kbps

Figure 13–4 is a typical public data network interconnect arrangement showing all three modes of operation. The packet assembler/disassembler (PAD) provides the necessary assembly/disassembly functions to interface non-packet-mode users to the public data network. Asynchronous users must comply with additional standards, such as the CCITT X.28 and X.29 to emulate point-to-point connections for nonintelligent terminals operating within the data network.

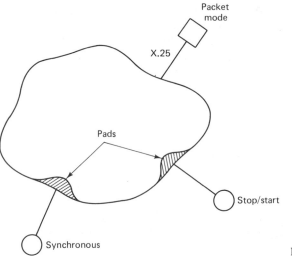

Figure 13–4   Public data network.

CCITT X.25: THE USER–NETWORK INTERFACE

In 1976, the CCITT defined the X.25 standard as the international standard for packet network access. The X.25 is strictly a user-to-network interface standard. X.25 addresses only the lower three levels of the ISO protocol hierarchy (the physical, data link, and network levels). Whenever possible, the X.25 uses existing standards. For example, for the physical level, X.25 specifies existing DTE/DCE standards. The CCITT X.21, X.26, and X.27 standards are specified as the physical interfaces for X.25. These standards correspond to EIA standards RS 232C, RS 423A, and RS 422A, respectively. For the second level, the data link level, X.25, defines the ISO HDLC as the international standard and the American National Standards Institute

(ANSI) 3.66, *Advanced Data Communications Control Procedures* (ADCCP), as the United States Standard. The data link level provides an envelope in which the data of the higher levels are embedded. Unfortunately, many of the features of these data link control procedures are unnecessary for packet networks. HDLC and ADCCP were designed for two-point private-line and multipoint polling applications. Consequently, the addressing and control procedures outlined by them are superfluous for packet networks. HDLC and ADCCP were selected because their block format and error detection mechanism are ideally suited to a packet environment. Level three, the network level, of the X.25 standard specifies three types of switching services offered in the packet mode on a switched data network. They are the *permanent virtual circuit,* the *virtual call,* and the *datagram.* All three modes transfer messages in packet format. The X.25 simply outlines the format the data must be in when they enter the data network.

Permanent Virtual Circuit (PVC).

The PVC is logically equivalent to a leased line, but slower. A PVC guarantees the connection, on demand, between two predetermined users of the network. Since the users of the PVC are fixed, there is no need to include identification codes or addresses when initiating a call or transferring data.

Virtual Call (VC).

The VC is logically equivalent to a dial-up circuit using the public telephone network. The VC provides a one-to-many access capability. Each user of the network has access to every other user through a common network of switches. Properties essential to a VC are sequenced data transfer, transparency, full-duplex capabilities, flow control, error control, and interface independence.

Datagram (DG).

The datagram is not completely defined in the X.25 standard. Currently, work is being done to standardize the datagram concept. Until agreed upon, its usefulness is questionable. A DG is the functional opposite of a VC. Users send small messages into the network. Each message is an independent entity and is not acknowledged. The network does not guarantee that the messages will arrive at the destination in the same order in which they were sent. However, very often the message will fit into a single packet. This is referred to as a *single-packet-per-segment protocol.* The DG is the most unreliable of the three packet modes of service.

X.25 packet formats.

The switched virtual call makes the most efficient use of the packet network. The virtual call has two packet frame formats, the call-request frame and the data-transfer frame.

Call-request frame. Figure 13–5 shows the field structure of the call-request frame. Each frame is delimited by a beginning and an ending flag. Following the beginning flag is the link address field, then a link control field. A CRC-16 frame check sequence precedes the ending flag. The flag, address, control, and frame check sequence fields are defined by the ISO's link-level protocol, HDLC.

1. *Flag field:* The flag sequence is 01111110. The flag identifies the beginning and ending of the packet.

Flag	Link address field	Link control field	Format identifier	Logical channel identifier	Packet type	Calling address length	Called address length	Called address
Bits: 8	8	8	4	12	8	4	4	To 60

Calling address	0	Facilities field length	Facilities field	Protocol ID	User data	Frame check sequence	Flag
Bits: To 60	2	6	To 512	32	To 96	16	8

Figure 13-5 Call-request packet frame format.

2. *Address field:*   The address field contains the link-level address. Either the basic 8-bit or extended $8 + 8N$-bit addressing modes may be used.

3. *Control field:*   This field is used for local counting and error control and correction between the DCE and the DTE. Either the basic 8-bit or the extended 16-bit control format may be used.

4. *Format identifier:*   The format identifier defines the nature of the packet—new call request, data packet, or a previously established call. The format identifier also identifies the numbering sequence limitations used (either 8 or 128).

5. *Logical channel identifier (LCI):*   The logical channel identifier designates a 12-bit logical channel number for the packet. Up to 4096 logical channels may be in use at any one time. The LCI, along with the called and calling addresses, is used to identify the source user and the destination user. When a user gains access to the network, it is assigned an LCI. When a user disconnects from the network, it relinquishes the assigned LCI. Once a call has been established, data transfer packets need only to include the LCI and not the calling or called addresses.

6. *Packet type:*   This 8-bit sequence identifies the function and the content of the packet—new request, call clear, call reset, and so on. This field also identifies flow control and data restrictions.

7. *Calling address length:*   This 4-bit field identifies (in binary) the number of digits included in the address of the calling party. Fifteen is the largest number of digits that may be used.

8. *Called address length:*   Same as the calling address length except that it identifies the number of digits included in the address of the called party. Again, the maximum number of digits that may be used is 15.

9. *Called address:*   This field contains the destination's address in binary-coded decimal (BCD) notation. The number of BCD codes in this field was previously specified in the called address length field. Each BCD number is made up of 4 bits. Fifteen 4-bit BCD codes represent a maximum of 60 bits for this field.

10. *Calling address:*   Same as the called address field except that this field contains the address of the calling party.

11. *Facilities length field:*   This field identifies, in binary, the number of 8-bit octets included in the following facilities field.

12. *Facilities field:*   This field contains information concerning optional network facilities, such as reverse billing, a closed user group, one-way transmit connections, and one-way receive connections.

13. *Protocol identifier:*   This 32-bit field can be used by the subscriber for unique, user-level protocol functions, such as log-on procedures and user identification practices.

14. *User data field:*   Up to 96 bits of user information can be transmitted with a data request packet. These are nonsequenced data and usually contain such information as the user's password number.

Data-transfer packet.   Figure 13–6 shows the field format for a data-transfer packet. It is similar to the call-request packet except that it contains considerably less overhead. The DLC frame format is HDLC. A data-transfer packet includes a send packet field and a receive packet field that were not included in the call-request packet.

1. *Send packet sequence number:*   This field contains a sequential numbering scheme used to identify packets within a logical channel identifier. The $P(s)$ count is used essentially the same way as the Ns is used in HDLC and SDLC. Each successive packet transmitted is assigned the next $P(s)$ number in the sequence. The $P(s)$ can be a 3-bit or a 7-bit binary number. Consequently, the maximum number of unacknowledged packets that may be outstanding is either 7 or 127. The number of $P(s)$ bits used is identified in the format identifier field. The send packet field is 8 bits long; unused bits are reset to a logic 0 condition.

2. *Receive packet sequence number:*   This field contains the number $P(r)$ of the last packet received on this LCI without error. The $P(r)$ is used to ACK/NAK previously received data packets.

Flag	Link address field	Link control field	Format identifier	Logical channel identifier	Send packet sequence number P(s)	0	Receive packet sequence number P(r)	0	User data	Frame check sequence	Flag
Bits: 8	8	8	4	12	3/7	5/1	3/7	5/1	To 1024	16	8

Figure 13–6   Data-transfer packet frame format.

LOCAL AREA NETWORKS

The decrease in the price of computers, word processors, high-speed terminals, and so on, has allowed businesses to install this equipment in numerous offices. Different types of computers are purchased to satisfy different goals. Local area networks (LANs) are designed to interconnect this equipment, with the obvious advantages of resource sharing and simplified managerial controls. LANs are privately owned and operated and are generally used to connect stations in the same building or in buildings in the same locale. Most LANs are 1 to 2 miles in length. A gateway is provided by which this network connects to the outside world. This eliminates the problem of interfacing various equipment with different networks.

Network Considerations

Topology.   The topology or architecture of a network identifies how the various stations will be interconnected. The star, bus, ring, and mesh topologies are the most common, each with its own advantages and disadvantages. These topologies are shown in Figure 13–7.

Connecting medium.   Although communications may be established on twisted pairs, coaxial cable, and fiber optics, all present local area network systems use coaxial cable as the transmission medium.

Information transmission.   The question now arises as to how information will be placed on this cable. There are two basic approaches: baseband and broadband transmission. *Baseband* transmission uses the cable as a *single channel* device. All

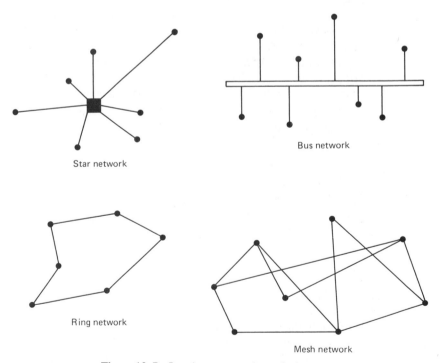

Figure 13–7   Local area network topologies.

stations connected to this channel must transmit the same type of information at the same speed. *Broadband* transmission is a *multichannel* operation. Each channel occupies a different frequency band and the information on each channel may be multiplexed. This permits voice, digital data, and video information to be transmitted simultaneously. Because of the need of modems, amplifiers, and more complex transmitters and receivers, a broadband system is more complex than a baseband system.

Channel access.   Channel access defines how a station gets control of the channel to transmit. Dedicated channels have been discounted because of high inefficiency due to nonuse of the channel for a large portion of the time. Contrasted with dedicated channels is random access. Although many means are available for achieving random access, *carrier sense, multiple access with collision detection (CSMA/CD)* and *token passing* have been receiving the most attention. With CSMA/CD, a station "listens" to detect if any transmissions are currently being sent by other stations. If so, that station will wait until the transmission ends and then transmit. If another station started to transmit at the same time, a collision of data will occur. In such a case, both stations will immediately stop transmitting and back-off for a random period of time before attempting to retransmit. A detailed explanation of CSMA/CD as used by Ethernet is given in a subsequent discussion. CSMA/CD is used by most baseband networks. Token passing is suited for a ring topology network. An electrical token is placed on the channel. As it circulates around the ring from station to station, if a particular station wishes to transmit, it removes the token, places its message on the channel, then reinserts the token after its message. The station possessing the token is the only one permitted to transmit. As the message circulates, it is removed at its intended destination. With token passing, each station is assured of having access to the channel in turn and of being able to transmit all of its messages. This guarantee of successful message transmission is not available with CSMA/CD.

Systems

This is not intended to be an inclusive list of all existing LANs. The networks shown demonstrate the various channel accesses used with either baseband or broadband networks.

Ethernet:	developed by Xerox Corporation in conjunction with Digital Equipment Corporation and Intel Corporation. It is a baseband system using CSMA/CD.
Wangnet:	developed by Wang Computers. It uses CSMA/CD with a broadband network.
Localnet:	developed by Sytek. It also uses CSMA/CD with a broadband network.
Domain:	developed by Apollo Computers. It uses a token-passing ring structure in a broadband network.

IEEE 802 Local Area Network Committee

Although the same accessing technique may be used by different manufacturers, this does not mean that all used the same codes, signal levels, and so on. In 1980, the 802 Local Area Network Committee was established to standardize the means of connecting digital computer equipment and peripherals with the local environment. In 1983, the IEEE standards board approved standard 802.3—CSMA/CD accessing on a bus topology—and standard 802.4—token passing accessing on a bus topology. These standards, together with standard 802.2, which describes the logical link control, have been adopted by ISO. Included in the standards that are still pending is the 802.5, which describes token passing accessing in a ring topology. These standards are expected to be approved in 1984.

ETHERNET

Ethernet is a baseband system which uses a bus topology and is designed for use in a local area network. Since there are many computer manufacturers, the problem is to design a system in which different types of computers can communicate with each other. In a demonstration in 1982, the computers of 10 major companies were linked on a 1500-ft Ethernet cable for electronic mail, word processing, and so on.

Before explaining the operation of Ethernet, a description of this system's hardware and software is provided.

Software

Information is transmitted from one station (computer) to another in the form of packets. The format for a packet is shown in Figure 13–8. The preamble is used for bit synchronization. The source and destination addresses and the field type make up the header information. The CRC is computed on the header information and the data field. The packet information is converted to Manchester code and transmitted at a 10-Mbps rate. In the Manchester encoding, each bit cell is divided into two parts. The first half contains the complement of the bit value and the second contains the actual bit value. This is illustrated in Figure 13–9. This code ensures that a signal transition exists in every bit.

The destination of a packet may be a single station or a group of stations. A *physical address* is a unique address of a single station and is denoted by a 0 as the first

Preamble	Destination address	Source address	Type field	Data field	CRC
8	6	6	2	46–1500	4

Bytes:

Figure 13–8   Packet format for Ethernet.

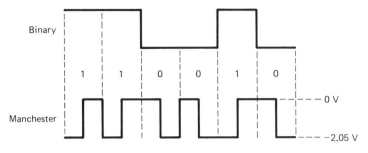

Figure 13–9   Manchester code.

bit (LSB) of the destination address. A *multicast address* has two forms. If only a partial group of the total stations is addressed, the first bit is a 1. The destination address is all 1's if all of the stations in the network are to receive the transmitted packet. A delay of 9.6 μs is required between the transmission of packets.

Hardware

Coaxial cable is the transmission medium for Ethernet. A cable segment may have a maximum distance of 500 m (1640 ft). Each segment may have up to 100 transceivers attached by way of pressure taps. Since these taps do not require cable cutting for installation, additional stations may be added without interfering with normal system operations. Interlan has developed the NT10 transceiver, in which the taps consist of two probes. One probe contacts the center conductor while the other contacts the shield. Color bands on the segment cable identify where the taps may be placed. A maximum of three segments may be connected end to end by way of repeaters to extend the system length to 1500 m (4921 ft). In Figure 13–10, this is the path between A and C. Remote repeater may be used with a maximum distance of 1000m between them. If these are used, the maximum end-to-end distance would be extended to 2.5 km (8202 ft). In Figure 13–10, if the center and right segments were interchanged, there would be a 2.5-km separation between A and B. There must always be only one signal path between any two stations. A station is connected to a controller, which is then interfaced with the transceiver to the transmission system. This is shown in Figure 13–11.

System Operation

Transmission.   The data link control for Ethernet is CSMA/CD. Stations acquire access to the transmission system through contention. They are not polled, nor do they have specific time slots for transmission. A station wishing to transmit first determines if another station is currently using the transmission system. The controller accomplishes this through the transceiver by sensing the presence of a carrier on the line. The controller may be a hardware or a software function, depending on the complexity of the station. In 1983, Xerox Corporation, in conjunction with Digital Equipment Corporation and Intel Corporation, developed and introduced a single-chip

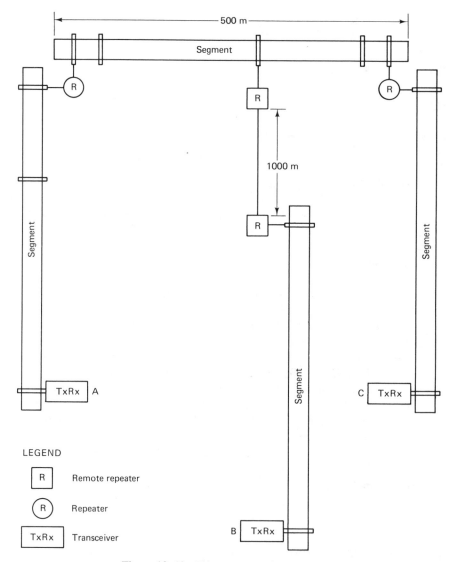

Figure 13-10   Ethernet transmission system.

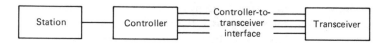

Figure 13-11   Ethernet station connection to the transmission system.

controller for Ethernet. Later, Mostek Corporation, working with Digital Equipment Corporation and Advanced Micro Devices, announced a two-chip set for this purpose. One chip is the local area network controller for Ethernet (LANCE) and the second is a Serial Interface Adapter.

The presence of a carrier is denoted by the signal transitions on the line produced by the Manchester code. If a carrier is detected, the station defers transmission until the line is quiet. After the required delay, the station sends digital data to the controller. The controller converts these data to Manchester code, inserts the CRC, adds the preamble, and places the packet on-line. The transmission of the entire packet is not yet assured. A different station may also have detected the quiet line and started to transmit its own packet. The first station monitors the line for a period called the *collision window* or the *collision interval*. This interval is a function of the end-to-end propagation delay of the line. This delay, including the delay caused by any repeaters, measured in distance, cannot exceed 2.5 km. If a data collision has not occurred in this interval, the station is said to have line acquisition and will continue to transmit the entire packet. Should a collision be detected, both of the transmitting stations will immediately abort the transmission for a random period of time and then will attempt to retransmit. A data collision may be detected by the transceiver by comparing the received signal with the transmitted signal. To make this comparison, a station must still be transmitting its packet while a previously transmitted signal has propagated to the end of the line and back. This dictates that a packet be of some minimum size. If a collision is detected, it is the controller that must take the necessary action. The controller-transceiver interface contains a line for notification of Collision Presence (10-MHz square wave). The remaining three lines of this interface are the Transmit Data, Receive Data, and power for the transceiver. Since a collision is manifested in some form of phase violation, the controller alone may detect a collision. In Ethernet, data collision is detected mainly by the transceiver. Where feasible, this is supplemented by a collision detection facility in the controller. To ensure that all stations are aware of the collision, a *collision enforcement consensus procedure* is invoked. When the controller detects a collision, it transmits four to six bytes of random data. These bytes are called the *jam sequence*. If a collision has occurred, the station's random waiting time before retransmission is determined from a *binary exponential back-off algorithm*. The transmission time slot is usually set to be slightly longer than the round-trip time of the channel. The delay time is randomly selected from this interval.

Example

Transmission time slot = time of 512 bits

$$\text{Maximum time of the interval} = \frac{512 \text{ bits}}{10 \text{ Mbps}} = 51.2 \ \mu s$$

Time interval = 0 − 51.2 μs

For each succeeding collision encountered by the same packet, the time interval is doubled until a maximum interval is reached. The maximum interval is given as

$2^{10} \times$ transmission time slot. After 15 unsuccessful attempts at transmitting a packet have been made, no further attempts are made and the error is reported to the station. This is the major drawback of Ethernet—it cannot guarantee packet delivery at a time of heavy transmission load.

Reception. The line is monitored until the station's address is detected. The controller strips the preamble, checks the CRC, and converts the Manchester code back to digital format. If the packet contains any errors, it is discarded. The end of the packet is recognized by the absence of a carrier on the transmission line. This means that no transitions were detected for the period of 75 to 125 ns since the center of the last bit cell. The decoding is accomplished through a phase-locked loop. The phase-locked loop is initialized by the known pattern of the preamble.

QUESTIONS

1. If there were 14 terminals in a network:
   (a) How many direct connections would be required to hook up every terminal to every other terminal?
   (b) How many connections would be required if a switch was to interconnect any two terminals?
2. Select answers from the following: (a) circuit switching; (b) message switching; (c) packet switching. Which switching technique(s):
   (1) Operate in real time or near real time?
   (2) Are transparent?
   (3) Are transactional?
   (4) May encounter blocking?
   (5) Require end terminals to be compatible?
   (6) Are a form of hold-and-forward network?
   (7) Are a form of store-and-forward network?
   (8) Transmit all information in the smallest segments?
3. HDLC is the standard protocol for packet switching.   (T, F)
4. All seven levels of ISO protocols are defined also by CCITT X.25.   (T, F)
5. CCITT X.25 describes the standards between two switches.   (T, F)
6. CCITT X.25 describes packet formats.   (T, F)
7. What is the approximate end-to-end distance of most local area networks?
8. If cable is used for only a single channel, that local area network is said to be using a _____ system.
9. Data collisions are possible with token passing.   (T, F)
10. Ethernet is a form of topology.   (T, F)
11. CSMA/CD is used only with baseband systems.   (T, F)
12. If an Ethernet system was extended from 500 m to 1000 m, the minimum packet size would (*have to be increased, have to be decreased, not have to be changed*).
13. On Ethernet, how is a physical address distinguished from a multicast address?

14. The Ethernet carrier is a constant-frequency signal.   (T, F)

15. Which piece of equipment normally detects collisions on Ethernet?

16. Identify three nonrelated functions performed by the controller.

17. If an Ethernet message experienced collisions 15 times, the interval from which a back-off time was selected was doubled each time.   (T, F)

18. Draw the Manchester-coded waveform for the following digital waveform:

INDEX